BASIC PRINCIPLES OF MARINE NAVIGATION

by

D. A. MOORE

U.K. Standard Book Number
85309 011 4

KANDY PUBLICATIONS LIMITED

PREFACE

Navigation can be achieved with a fair measure of success if one simply substitutes numerals for certain symbols in the various standard formulae. However, it is desirable that the navigator should know the principles involved as then he will be in a position to assess the reliability of his navigation.

While not probing the far depths, the aim of this book is to present to students for the M.O.T. examinations and for G.E.C. "O" level Navigation, the basic principles of this science as practiced at sea. It will be found that the astronomical navigation required by both syllabuses is adequately covered.

The text is concise but explicit and is illustrated by a large number of diagrams and worked examples. It commences with a section on trigonometry before gradually leading to sections dealing with the sailings, time, nautical astronomy, the taking and plotting of astronomical observations and an explanation of the tides. A number of revision questions are included and there is an adequate index.

I am indebted to H.M. Stationery Office, The Hydrographic Department of the Ministry of Defence and the Liverpool Tidal Institute and Observatory for permission to use extracts from the 1965 edition of the Admiralty Tide Tables which are Crown Copyright, also to Messrs. Imray, Laurie, Norie and Wilson for permission to reprint extracts from their A.B.C. tables.

My thanks are also due to those who have assisted in the preparation of this book, particularly with the checking of the manuscript and proof reading. While every endeavour has been made to ensure accuracy I should be pleased to hear of any errors which may have occurred.

Buckhurst Hill, Essex.
November 1964.

D.A. Moore

CONTENTS

I
INTRODUCTION

It is necessary, in a book of this small size, to assume that the reader has attained a certain standard in mathematics. However, the following mathematical formulae are appended as a memory refresher.

TRIGONOMETRICAL RATIOS:

In the right-angled triangle ABC right-angled at C

$$\sin\theta = \frac{a}{c} = \cos(90^\circ - \theta) = \frac{1}{\operatorname{cosec}\theta}$$

$$\cos\theta = \frac{b}{c} = \sin(90^\circ - \theta) = \frac{1}{\sec\theta}$$

$$\tan\theta = \frac{a}{b} = \cot(90^\circ - \theta) = \frac{1}{\cot\theta}$$

$$\operatorname{cosec}\theta = \frac{c}{a} = \sec(90^\circ - \theta) = \frac{1}{\sin\theta}$$

$$\sec\theta = \frac{c}{b} = \operatorname{cosec}(90^\circ - \theta) = \frac{1}{\cos\theta}$$

$$\cot\theta = \frac{b}{a} = \tan(90^\circ - \theta) = \frac{1}{\tan\theta}$$

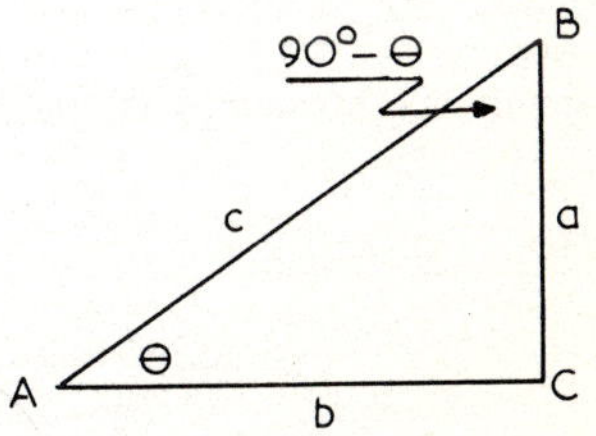

Angles are complementary when their sum = 90°
Angles are supplementary when their sum = 180°

SIGNS OF THE TRIGONOMETRICAL RATIOS:

From 0° to 90° all ratios are positive
From 90° to 180° only the sin. and cosec. are positive
From 180° to 270° only the tan. and cotan. are positive
From 270° to 360° only the cos. and sec. are positive

PLANE TRIG. FORMULAE:

Sine rule: $\dfrac{a}{\sin A} = \dfrac{b}{\sin B} = \dfrac{c}{\sin C}$

Cosine rule:

$$a^2 = b^2 + c^2 - 2.b.c.\cos A$$
$$b^2 = a^2 + c^2 - 2.a.c.\cos B$$
$$c^2 = a^2 + b^2 - 2.a.b.\cos C$$

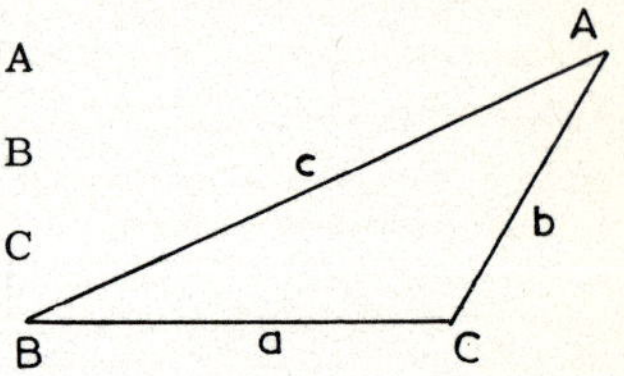

or: $\cos A = \dfrac{b^2 + c^2 - a^2}{2.b.c.}$

$$\cos B = \frac{a^2 + c^2 - b^2}{2.a.c.}$$

$$\cos C = \frac{a^2 + b^2 - c^2}{2.a.b.}$$

SPHERICAL TRIGONOMETRY:

A spherical triangle is bounded on all three sides by arcs of great circles, i.e. circles whose planes pass through the centre of the sphere. All sides and angles of spherical triangles are measured in angular measure.

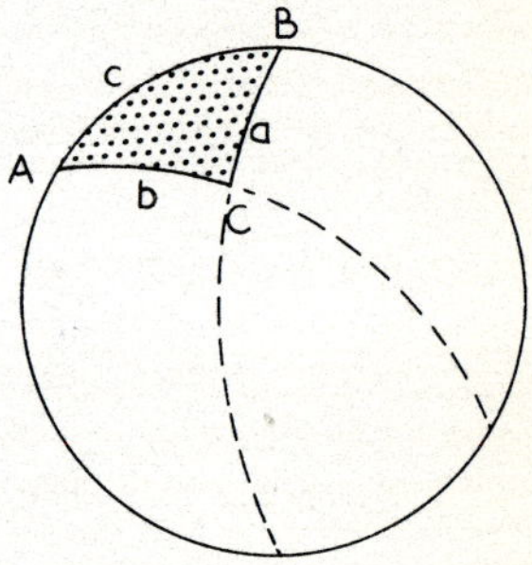

THE SINE RULE

$$\frac{\sin a}{\sin A} = \frac{\sin b}{\sin B} = \frac{\sin c}{\sin C}$$

THE COSINE RULE

$$\cos a = \cos b.\ \cos c + \sin b.\ \sin c.\ \cos A.$$
$$\cos b = \cos a.\ \cos c + \sin a.\ \sin c.\ \cos B.$$
$$\cos c = \cos a.\ \cos b + \sin a.\ \sin b.\ \cos C.$$

The cosine rule is unsuitable for logarithmic work because the cosines of angles in the second quadrant have negative values. To overcome this difficulty, the haversine of the angle is used. This function is half the versine, which is defined as the difference between the cosine of the angle and unity.

$$\text{versine } \theta = 1 - \text{cosine } \theta$$
$$\text{haversine } \theta = \tfrac{1}{2}(1 - \text{cosine } \theta)$$

and re-arranging, $\cos \theta = 1 - 2.\text{hav } \theta$

The figure shows the haversine curve in relation to the cosine curve from which it is derived.

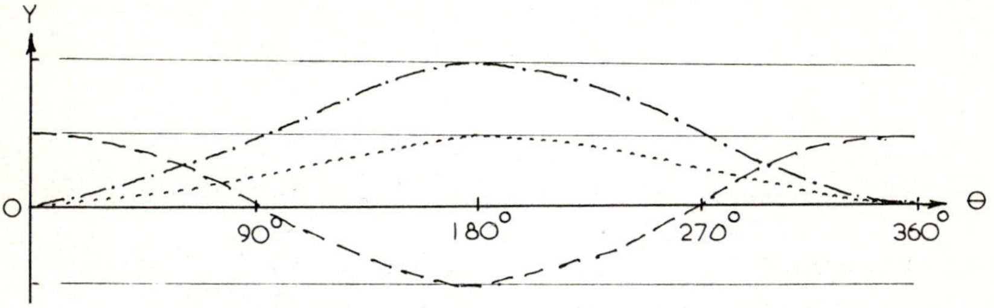

DERIVATION OF THE HAVERSINE FORMULA

$$\cos a = \cos b.\ \cos c + \sin b.\ \sin c.\ \cos A.$$

and substituting for cos A

$$\cos a = \cos b.\ \cos c + \sin b.\ \sin c. (1 - 2.\,\text{hav}\ A)$$

i.e. $\cos a = \cos b.\ \cos c + \sin b.\ \sin c - 2.\sin b.\ \sin c.\ \text{hav}\ A.$

$= \cos (b \sim c) - 2.\sin b.\ \sin c.\ \text{hav}\ A.$

Note:- $\cos (b + c) = \cos b.\ \cos c - \sin b.\ \sin c.)$
$\cos (b - c) = \cos b.\ \cos c + \sin b.\ \sin c.)$

and substituting for cos a and cos $(b \sim c)$

$$1 - 2.\text{hav}\ a = 1 - 2.\text{hav}\ (b \sim c) - 2.\sin b.\sin c.\text{hav}\ A.$$

i.e. $\text{hav}\ a = \text{hav}\ (b \sim c) + \sin b.\sin c.\text{hav}\ A.$

All quantities are positive and the formula is suitable for logarithmic work. The term (sin b. sin c. hav A) will always be positive and less than unity, and this simplifies the use of the formula because the antilogarithm of the term can be found immediately from the tables where natural and logarithmic values of the haversine are printed side by side.

EXAMPLE 1

It is required to find b when:- $a = 40^\circ$ $c = 75^\circ$ $B = 56^\circ$

$$\text{hav}\ b = \text{hav}\ (a \sim c) + \sin a.\sin c.\text{hav}\ B$$

log hav B	9.34322	$B = 56^\circ$
log sin c	9.98494	$c = 75^\circ$
log sin a	9.80807	$a = 40^\circ$
log sin a.sin c.hav B	9.13623	
nat. sin a.sin c.hav B	0.13684	
nat.hav $(a \sim c)$	0.09042	$(a \sim c) = 35^\circ$
nat.hav b	0.22726	

and from the tables $b = 56^\circ\ 56'.6$

DERIVATION OF THE FOUR-PART FORMULA

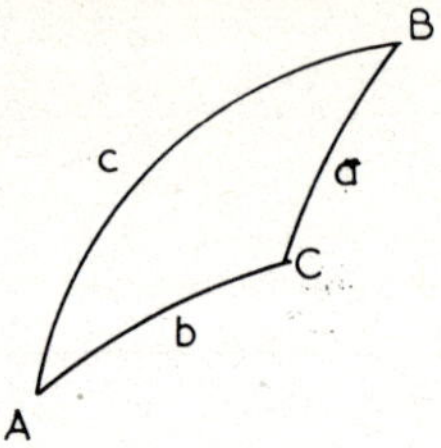

In the above triangle

$$\cos b = \cos a.\cos c + \sin a.\sin c.\cos B \quad \text{----------(1)}$$
$$\cos a = \cos b.\cos c + \sin b.\sin c.\cos A \quad \text{----------(2)}$$

$$\sin a = \frac{\sin b.\sin A}{\sin B} \quad \text{-----------------------------(3)}$$

substituting (2) for cos a, and (3) for sin a in (1)

$$\cos b = \cos b.\cos^2 c + \cos c.\sin b.\sin c.\cos A + \frac{\sin b.\sin A}{\sin B}\sin c.\cos B$$

re-arranging:-

$$\cos b - \cos b.\cos^2 c = \cos c.\sin b.\sin c.\cos A + \sin b.\sin A.\sin c.\cot B$$

or: $$\cos b.\sin^2 c = \cos c.\sin b.\sin c.\cos A + \sin b.\sin A.\sin c.\cot B$$

note $\cos b - \cos b.\cos^2 c. = \cos b\,(1 - \cos^2 c)$
$= \cos b.\sin^2 c$ Since $(1 - \cos^2 c) = \sin^2 c$

dividing throughout by sin b. sin c:

$$\frac{\cos b.\sin^2 c}{\sin b.\sin c} = \cos c.\cos A + \sin A.\cot B$$

thus re-arranging:-

$$\cot b.\sin c = \cos c.\cos A + \sin A.\cot B$$

This is the four-part formula whereby, if three of four adjacent parts of a spherical triangle are known, the fourth may be found.

A useful aide memoire for the four part formula is given with the figure below.

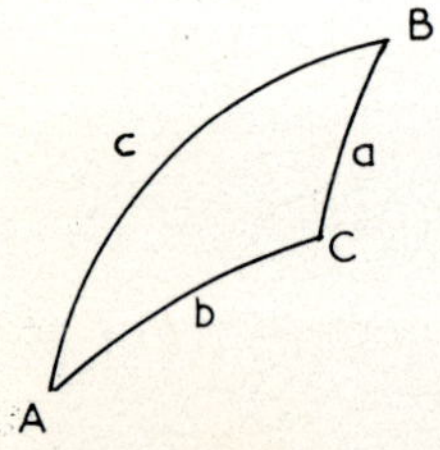

B, c, A, b, are the four adjacent parts and

B - outer angle OA
A - inner angle IA
b - outer side OS
c - inner side IS

cos IS cos IA = sin IS cot OS - sin IA cot OA

RIGHT ANGLED SPHERICAL TRIANGLES

These are best solved by the use of Napier's rules for circular parts.

It is essential to know two parts of the triangle, other than the right angle. To formulate the rules, the complements are taken of the two angles and included side opposite the right angle.

sine of the middle part = product of the tangents of the adjacent parts.
sine of the middle part = product of the cosines of the opposite parts.

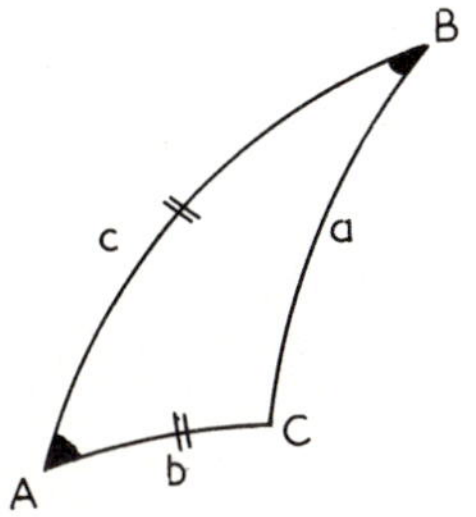

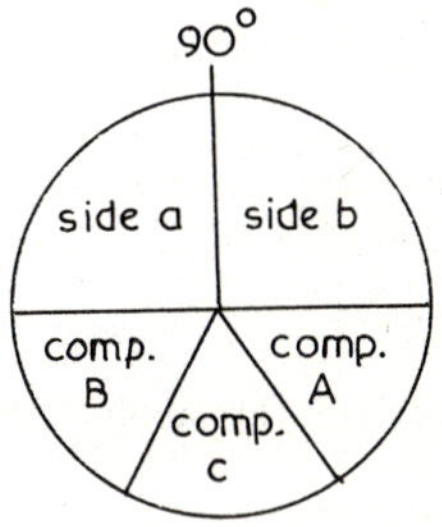

in the diagram above; given that $C = 90^{o}$ and given A and b:

to find B:

sin co. B = cos b. cos co. A

therefore cos B = cos b. sin A.

to find a:

sin b = tan a. tan co. A

sin b = tan a. cot A.

therefore tan a = sin b. tan A.

Remember:

Angles and sides opposite are always of like affection: i.e. both under or both over 90°.

When angles or sides are of the same affection, the hypotenuse is less than 90°.

When angles or sides are of contrary affection, the hypotenuse is greater than 90°.

When an angle and the side opposite are given, the triangle has two possible solutions; i.e. ambiguity exists.

One point must be taken note of in the solution of the quadrantal triangle; i.e. one that contains a side of 90°.

When finding the angle opposite the 90° side, or if given this angle and using it in the calculation, the algebraic sign of the angle must be changed.

II
TERRESTRIAL SPHERE

THE TERRESTRIAL SPHERE

The shape of the Earth is that of an oblate spheroid, which is the solid formed by rotating an ellipse about the minor axis. This gives rise to a flattening at the polar regions but this is so small that, for the purposes of practical navigation, the Earth could be considered to be a perfect sphere. The errors that arise from this assumption are usually negligible.

The Earth's movement in space has two detectable components:-

1. a daily rotation about a fixed axis.

2. a yearly passage in an elliptical path about the Sun.

It is the first of these that gives rise to the four cardinal directions.

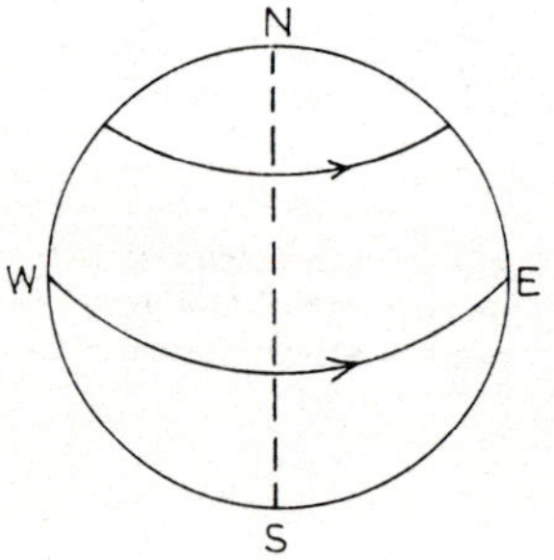

In the figure, NS is the diameter about which the Earth rotates once each day and the points N and S where this diameter meets the surface are known as the poles.

East is the direction in which any point on the Earth's surface is being carried by the rotation and is shown by the arrows. West is the direction 180° from east.

The poles are named arbitrarily, the north pole lying 90° to the left of an observer facing east and the south pole lying 90° to the right of the same observer.

Having obtained the four cardinal directions, the angle between each of these is further sub-divided into 8 parts called points, each having an angular measure of $11\frac{1}{4}^{o}$. They take their names from the two adjacent cardinal points.

Points				Angular Measure
N	E	S	W	0°
N by E	E by S	S by W	W by N	$11^{o}15'$
NNE	ESE	SSW	WNW	$22^{o}30'$
NE by N	SE by E	SW by S	NW by W	$33^{o}45'$
NE	SE	SW	NW	45°
NE by E	SE by S	SW by W	NW by N	$56^{o}15'$
ENE	SSE	WSW	NNW	$67^{o}30'$
E by N	S by E	W by S	N by W	$78^{o}45'$
E	S	W	N	90°

Nowadays, direction is given in the 360° notation, i.e. North is 000° or 360°, all other directions being measured clockwise from north.

The direction indicated by OA is 080°.
The direction indicated by OB is east or 090°.
The direction indicated by OC is 220°.

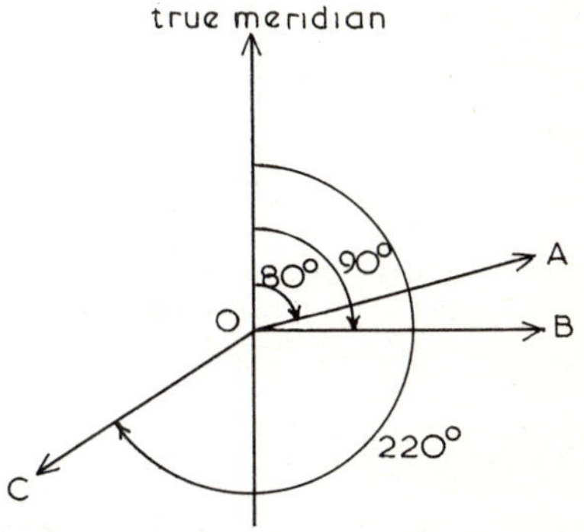

The navigator, at sea, has no indication of direction by merely surveying the horizon. He does, however, use the ship's compass which indicates the compass course.

The gyro-compass refers to the true meridian, and the course steered is then the angle the ship's head makes with the true geographical meridian, and is known as the true course.

The magnetic compass that is influenced only by the Earth's magnetic field indicates the magnetic meridian. Since the magnetic poles do not coincide with the geographic poles, there will always be some angle between the true and magnetic meridians (except when the true and magnetic poles are in line).

In the diagram alongside

N and S are the geographic north and south poles, respectively.
MN and MS are the magnetic north and south poles, respectively.
Angle N. O. MN is the angle between the true and magnetic meridians at O.

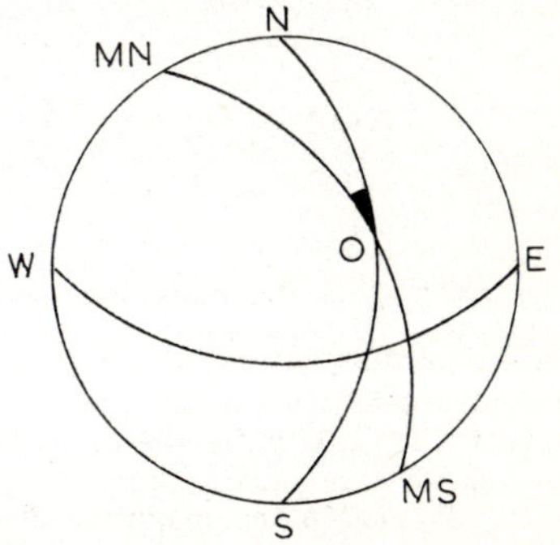

This angle between the two meridians is called variation (some physics textbooks refer to it as declination) and is named east when the magnetic meridian is angled to the right of the true meridian, and west when angled to the left.

The ship's head relative to the magnetic meridian indicates the magnetic course.

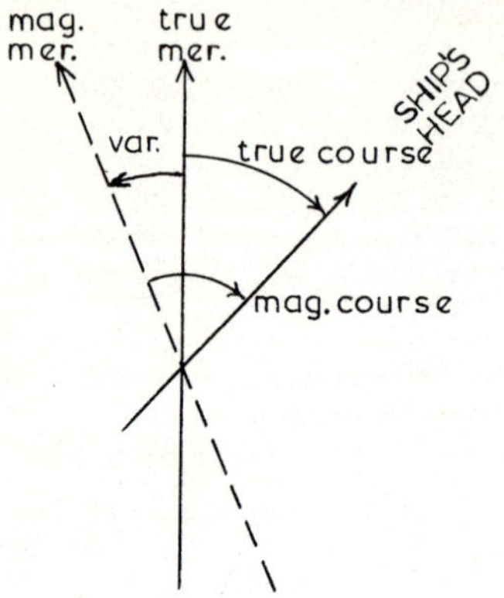

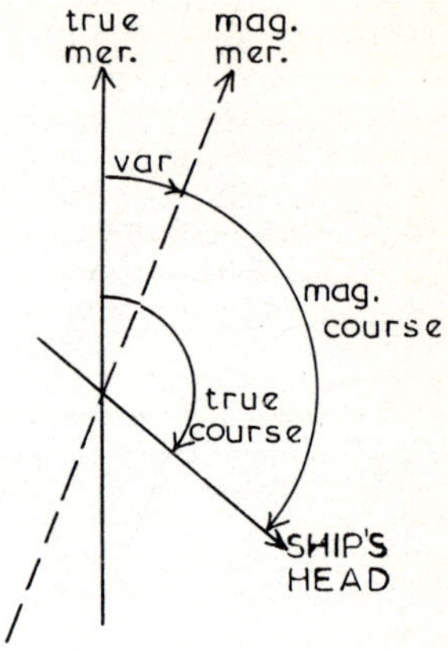

From the diagrams above it can be seen that to convert mag. courses to true courses, the variation (var) is added when east, and subtracted when west.

EXAMPLE 2

Convert 006° mag. course to true course when the variation is 16°E and 16°W.

magnetic course	006°M	006°M
variation	16°E	16°W
true course	022°T	350°T

EXAMPLE 3

Convert 273°T to magnetic when the variation is 23°E and 17°W.

true course	273°T	273°T
variation	23°E	17°W
magnetic course	250°M	290°M

Owing to the material used in the construction of ships, i. e. steel, all vessels tend to have some magnetic influence. The result of this is to deflect the compass needle from the magnetic meridian. This deflection is known as deviation and is named east when the needle is deflected to the right of the magnetic meridian, and west when it is deflected to the left.

To obtain the magnetic course, east deviation (dev) must be added to and west dev. must be subtracted from, the compass course.

The combination of var. and dev. is known as the error of the magnetic compass.

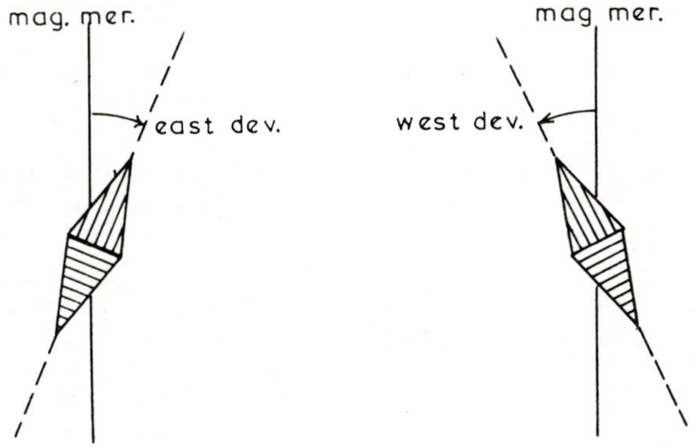

To obtain the amount and name of the error:

dev. and var. same names - add - error then has the same name.

dev. and var. different names - take the difference – error takes the name of the greater.

EXAMPLE 4

Ship's head by compass 003°C. dev. 15°E. var. 16°E. Find the true course.

ship's head	003°C.	or:	dev.	15°E.
dev.	15°E.		var.	16°E.
	018°M.		error	31°E.
var.	16°E.	ship's head		003°C.
ship's head	034°T.	ship's head		034°T.

EXAMPLE 5

True course 086°, var. 23°W. dev. 10°E. Obtain the error of the compass and the compass course to steer.

dev.	10°E.	ship's head	086°T.
var.	23°W.	error	13°W.
error	13°W.	ship's head	099°C.

Summarising:

compass to true - east error is added
C.E.R.T.

true to compass - east error is subtracted
T.E.L.C.

C.E.R.T. = compass - east error to right - true
T.E.L.C. = true - east error to left - compass

TERRESTRIAL POSITION

Before this is explained, two important definitions must be borne in mind:-

1. Great circle. Any section through a sphere results in the formation of a circle. If the plane of this circle passes through the centre of the sphere then the circle so formed is a great circle.

2. Small circle. All circles formed by sections, other than great circles are called small circles, i.e. their plane does not pass through the centre of the sphere.

In order to describe the position of a point on the Earth's surface, it is sufficient to know its shortest angular distance from two lines, perpendicular to each other, on the surface of the earth.

The two lines of reference are the Equator and the meridian of Greenwich. The Equator is a particular great circle whose plane is perpendicular to the Earth's axis of rotation. It divides the Earth into the northern and southern hemispheres, and is the datum line from which latitude is measured. The meridian of Greenwich is the semi-great circle, terminating at the poles, that passes through the Greenwich observatory and is the datum line from which longitude is measured.

All small circles, parallel to the equator, are called parallels of latitude and are named north or south according to the hemisphere in which they are situated. All semi-great circles terminating at the poles are called meridians of longitude and are named east or west according to whether they lie east or west of the Greenwich meridian. The continuation of the Greenwich meridian is the meridian of 180° long.

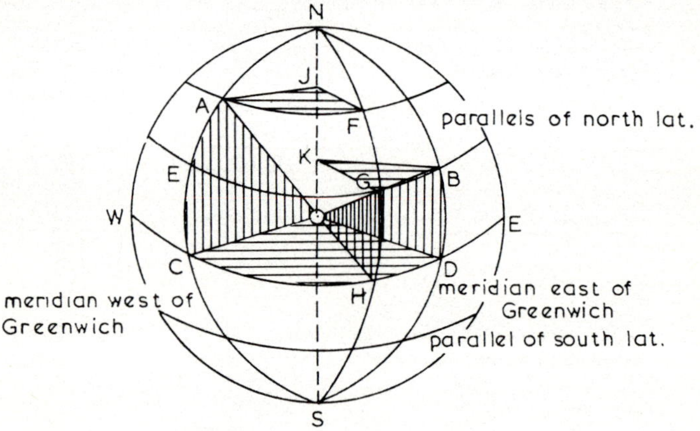

In the figure, A lies on a particular parallel of northern latitude and its latitude is angle COA. If NFGHS is the meridian of Greenwich then the west longitude of point A is the angle HOC.

Angle HOC = Angle FJA
= Angle HNC
= Arc HC

Similarly with point B, its latitude is angle DOB and its east longitude is angle GKB = arc HD. The difference of latitude (d. lat) between A and B is the arc of the meridian cut off by the parallels of latitude through A and B. (arc AE). The rule for finding the d'lat between two places may be summarised:-

Same names - subtract
Different names - add

The difference of longitude (d'long) between A and B is the shortest arc of the equator cut off by the meridians of A and B (arc CD), or the angle between the meridians at the pole (angle CND). The rule for finding the d'long is summarised:-

Same names - subtract
Different names - add
If result is greater than 180°, subtract from 360°.

One other method of describing terrestrial position is by means of a bearing and distance from a given or well known point. A ship, for example, may find a certain lighthouse bearing 225° distant 15 miles, conversely, the ship is bearing 045° distant 15 miles from the light. This method is only used when the relative distance between the points is small.

UNITS OF DISTANCE

The Statute Mile. This is an arbitrary unit of measure equal to 5280 feet and is never used in navigation.

The Geographical Mile. This is the length of an arc of the Equator subtending an angle of one minute at the centre of the Earth. The actual length of the geographic mile being 6087.2 feet.

The Nautical Mile. This is the length of an arc of a meridian subtending an angle of one minute at the centre of curvature of the place. Due to the oblateness of the Earth, the nautical mile increases as the latitude increases. The actual values are 6046.1 feet at Latitude 0° increasing to 6108.3 feet at the poles. For purposes of navigation, the convenience of having a fixed or standard unit is obvious, and the unit is the standard nautical mile of 6080 feet.

The navigational unit of speed is the knot and is one nautical mile per hour.

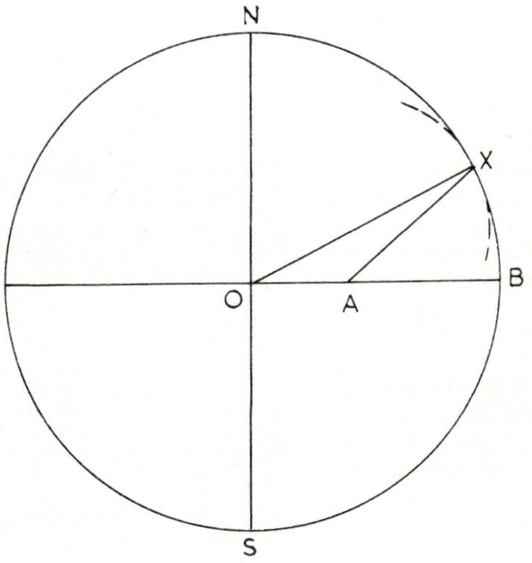

In the figure, O is the centre of the Earth
X is the place on the meridian NXS
A is the centre of curvature of the arc passing through X, i.e. the continuation of AX passes through the zenith of an observer at X.

The angle XAB is the geographical latitude of X and the angle XOB is the geocentric latitude of X.

III
RHUMB LINE SAILINGS

THE RHUMB LINE

A navigator, in sailing between two places, relies for direction upon his compass, be it magnetic or gyro, and consequently, steers a fixed course for a given period of time. This means that the ship's head will maintain a fixed angle with all the meridians that the ship crosses. This line of constant course is known as a rhumb line or loxodromic curve.

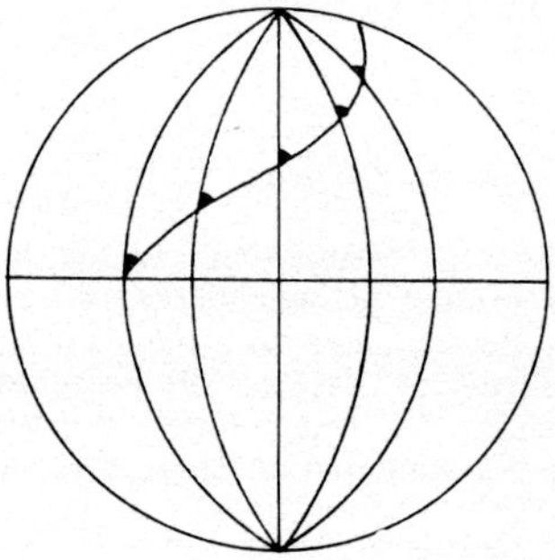

It can be seen from the diagram that the rhumb line does, in fact, spiral in towards the pole. Special cases of rhumb lines are the equator and parallels of latitude (constant course east or west) and the meridians (constant course north or south).

The sailings comprise various methods of calculating the course and distance between two places on the Earth's surface.

PARALLEL SAILING

A vessel on any course, other than north or south, moves some distance east or west. This is known as departure (dep) and is the arc of a parallel of latitude cut off by the meridians of the points of departure and arrival, measured in miles. Except at the equator, departure is always numerically less than the d'long expressed in minutes of arc.

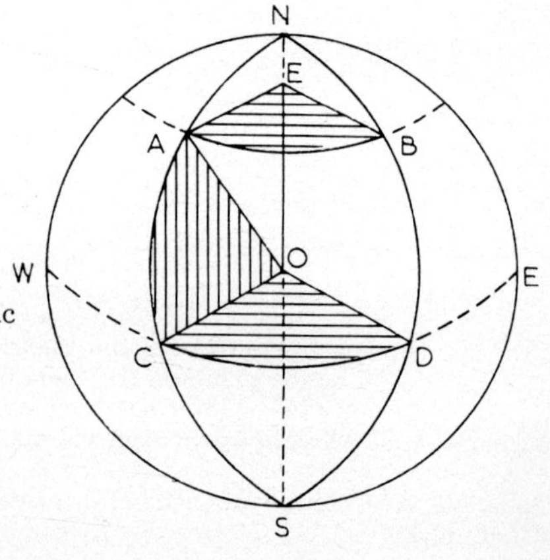

In the figure

AB = departure (miles)
CD = d'long (in minutes of arc which equals miles at equator)

$$\frac{\text{departure}}{\text{d'long}} = \frac{AB}{CD}$$

$$= \frac{AE}{CO}$$ (since arcs of concentric circles subtended by same angle are proportional to their radii)

$$= \frac{AE}{AO}$$

= cosine latitude.

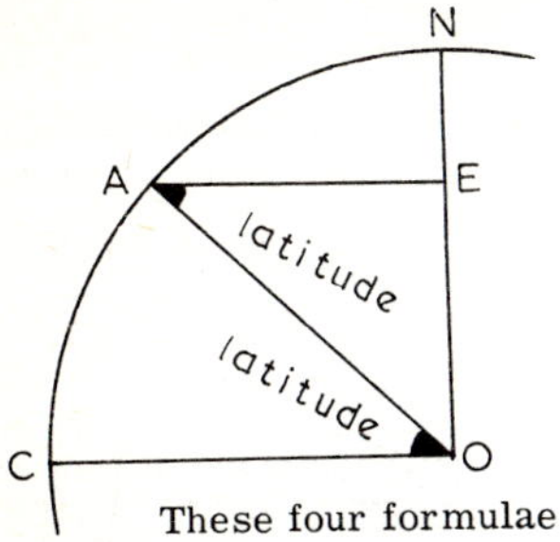

Therefore:-

$$\frac{\text{departure}}{\text{d'long}} = \text{cos. lat.}$$

$$\frac{\text{d'long}}{\text{departure}} = \text{sec. lat.}$$

$$\text{departure} = \text{d'long x cos. lat.}$$

$$\text{d'long} = \text{departure x sec. lat.}$$

These four formulae are collectively known as the parallel sailing formulae.

EXAMPLE 6

A vessel in position lat. $39^{\circ}00'$N., long. $40^{\circ}00'$W., steamed due east until her longitude was $35^{\circ}14'$W., How far did she steam?

long, left	$40^{\circ}00'$W.
Arr. long	$35^{\circ}14'$W.
d'long	$4^{\circ}46'$E = 286'E.

departure = d'long x cos. lat.
= 286' x cos. 39°
= 222.3 miles

Answer: Distance steamed due East = Departure = 222.3 miles

EXAMPLE 7

At what latitude is the d'long equal numerically to $2\frac{1}{2}$ times the departure?

$$\text{cos. lat} = \frac{\text{departure}}{\text{d'long}}$$

$$= \frac{\text{departure}}{\text{departure}} \times \frac{2}{5} \quad (\text{since d'long} = \frac{5}{2}\text{x dep.})$$

$$= \frac{2}{5} = 0.4$$

from the natural cos. tables the angle is $66^{\circ}25'$

Answer: Latitude $66^{\circ}25'$N. or S.

EXAMPLE 8

At what speed is a point on the Earth's surface in lat. 40° carried around the Earth's axis?

All points on the Earth's surface rotate through 360° in 24 hrs. Therefore in 1 hour the angular movement is 15°

d'long in 1 hour = 15° = 900'

departure = d'long x cos. lat.
= 900 x cos 40°
= 689.4 miles

Answer: Speed is 689.4 knots.

THE PLANE SAILING

In the figure below, a rhumb line is drawn between the departure and arrival points A and B. The angle between the ship's head and each meridian is constant and is known as the course angle.

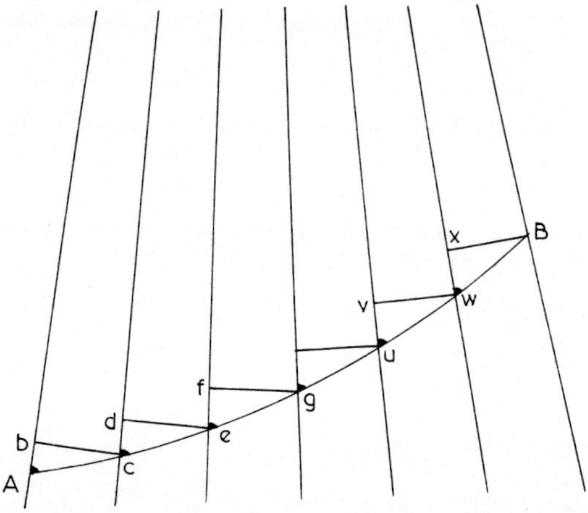

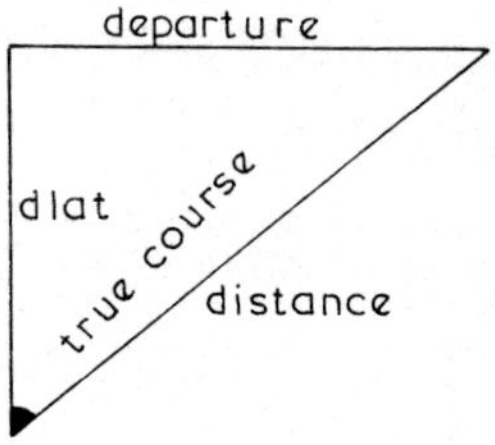

If the distance AB is divided into a large number of very small sections Ac, ce, eg, ... uw, wB. and if these sections are made small enough, then the triangles abc, cde, efg, ... uvw, wxB may be considered as plane triangles.

Sections Ab, cd, ef, ... uv, wx are sections of d'lat.

similarly bc, de, fg, ... vw, xB are sections of departure.

and by addition: bc + de + ... + vw + xB = departure

But:

bc = Ac sin. course
de = ce sin. course
fg = eg sin. course .... etc.

and by substitution:

dep = (Ac + ce + ... + uw + wB) sin. course

i. e. departure = distance x sine course. ..... (1)

Similarly it can be shown that:

d'lat = distance x cosine course. .... (2)

and dividing (1) by (2) :

$$\frac{\text{departure}}{\text{d'lat}} = \text{tangent course.}$$

Summarising the plane sailing formulae:

departure = distance x sin. course

d'lat = distance x cos. course

$$\frac{\text{departure}}{\text{d'lat}} = \text{tan. course.}$$

THE TRAVERSE TABLE

This table, which is included in all navigational tables, enables the navigator to solve any right angled triangle, in particular the plane sailing triangle, merely by inspection. It tabulates departure and d'lat for distances up to 600 miles for any course angle up to 90°.

An additional use of this table is for conversion of departure into d'long and vice versa. In this case the table is entered with the mean latitude. However, it must be borne in mind that a more accurate means of conversion of departure into d'long lies in the Parallel and Middle latitude sailings.

EXAMPLE 9

A vessel left a position with a light in lat. 43°25'N., long. 65°30'W bearing 062° distant 12 miles. Course set 210°T., log set at 0. With the log reading 14, course was altered to 300°T. and when it registered 32, course was again altered to 223°T. Find the course and distance made good when the log was reading 49, and the D.R. latitude.

	Dlat		Dep.	
	N.	S.	E.	W.
1st course 210° x 14'	-	12.1	-	7.0
2nd course 300° x 18'	9.0	-	-	15.6
3rd course 223° x 17'	-	12.4	-	11.6
	9.0	24.5S 9.0N		34.2W
Dlat and dep for course and dist.		15.5S		34.2W
Bearing reversed 242° x 12'		5.6S	not required	
Dlat for D.R. latitude		21.1S		

Latitude of light	43°25'.0N
d'lat	21'.1S
D.R. latitude	43°03'.9N

From Traverse Table and interpolating as necessary.
dlat 15'.5S dep. 34'.2W

gives course made good S65½°W = 245½°
distance made good 37.5 miles

THE MERCATOR SAILING

Before discussing this important sailing, it is essential to outline the theory of the Mercator chart.

It is necessary for the navigator to have maps of the Earth's surface in order to lay off his proposed course and to fix the position of his ship in relation to the adjacent land. This necessitates a flat drawing of part of the surface of a sphere. Distortion becomes inevitable, since a sphere cannot be unrolled onto a plane surface. The larger the area, the greater the distortion.

The most useful chart to the navigator is the Mercator chart, since he can draw a straight line between departure point and destination and so measure the steady course to be steered. The construction of the chart is such that:

1. rhumb lines on the Earth appear as straight lines on the chart.

2. the angles between rhumb lines are unaltered between the Earth and chart.

It must follow therefore:

1. the equator, a rhumb line as well as a great circle, will appear as a straight line.

2. parallels of latitude will appear as straight lines parallel to the equator.

3. the meridians will appear as straight lines perpendicular to the equator.

SCALE ON THE MERCATOR CHART

Since the equator is represented on the chart as a straight line of definite length, then the longitude scale is fixed by that length and is constant in all latitudes.

If the scale on a particular chart is 'a' inches to 1' of longitude and as previously shown,

$$\text{departure} = \text{a'long} \times \cos.\ \text{lat.}$$

$$\text{then departure (inches)} = 1'\ \cos.\ \text{lat.}$$

therefore one mile on this chart is represented by a. sec lat inches.

It follows, therefore, that the scale of latitude and distance on a Mercator chart is proportional to the secant of the latitude of that place. The amount of distortion is also governed by the secant of the latitude. Hence, Greenland, in 70°N. appears as broad as Africa at the equator, although, in fact, Africa is three times as broad as Greenland (sec $70^{o} = 3$).

MERIDIONAL PARTS.

Since the latitude and distance scale on a Mercator chart is proportional to the secant of the latitude, the scale increases as it recedes from the equator. At the poles (lat 90°) the value of the secant is infinity, and for this reason the polar regions cannot be shown on a Mercator chart. The latitude scale affords no ready means of comparison with the fixed longitude scale.

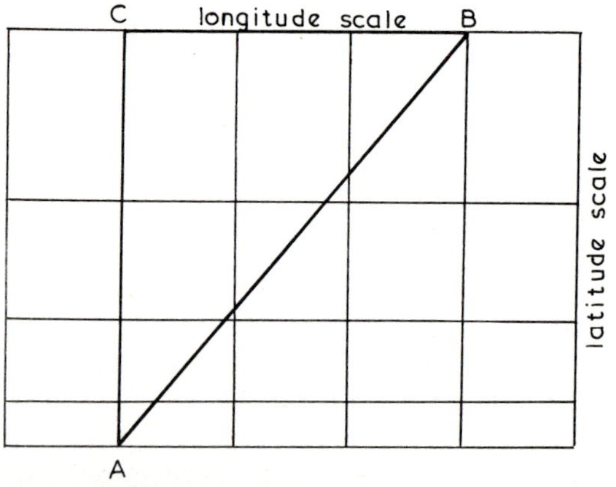

In the diagram, representing a rhumb line course AB drawn on a section of a Mercator chart, the tangent of the course is not BC (measured on the longitude scale) divided by AC (measured on the latitude scale). For this ratio to be true, AC and BC must be measured in the same fixed units. The longitude scale provides this unit, which is the length of one minute of arc on that scale. This length is known as a meridional part.

The meridional parts (mer. parts) of any latitude are the number of longitude units in the length of a meridian between the Equator and that parallel of latitude.

The mer. parts for any latitude are tabulated in all navigational tables, and since the Earth is a spheroid then the mer.parts for the spheroid or ellipsoid must be used. If the mer.parts for the sphere are used then the given latitude must be reduced in order to obtain the correct mer. parts.

The mer. parts (spheroid) for 45° are 3013.4 and if the longitude scale is such that one inch represents one degree (60 mer. parts), then the length of the meridian from the equator to the parallel of 45° is:

(3013.4 divided by 60) inches = 50.2 inches.

It can be seen that mer. parts involve chart lengths and are not connected in any way with distance on the Earth's surface, which is measured in nautical miles.

DIFFERENCE OF MERIDIONAL PARTS (d. m. p.)

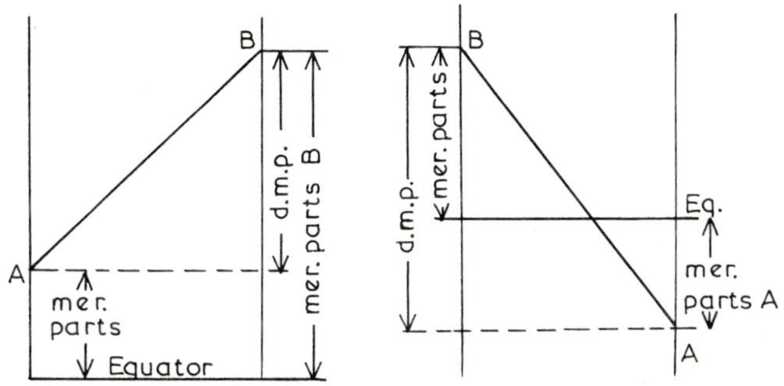

The number of mer. parts in the meridian between the parallels through A and B:

A and B on same side of equator: mer. parts B - mer. parts A.

A and B on opposite sides of the equator: mer.parts. A + mer. parts B.

This difference in mer. parts is always written as d. m. p.

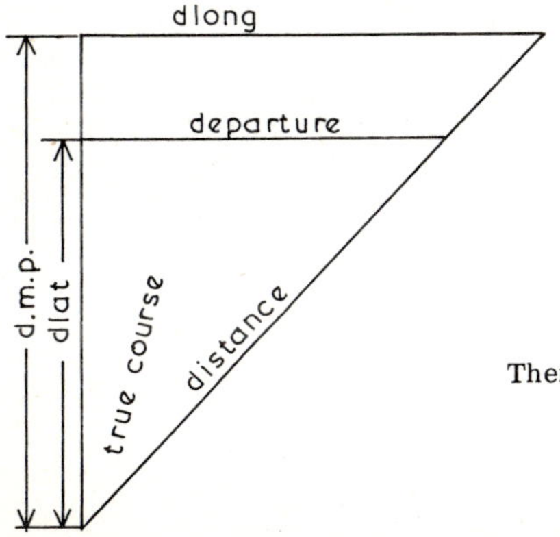

In this figure, the Mercator and plane sailing triangles are superimposed.

d'long and d. m. p. are in the same units.

Then: $\frac{\text{d'long}}{\text{d. m. p.}} = \frac{\text{departure}}{\text{d'lat}} = \text{tan. course}$

EXAMPLE 10

Find the course and distance between position A (lat. 16°00'S., long. 5°55'W.) and position B (lat. 23°35'S., long. 24°30'W.)

latitude A.	16°00'S	mer. parts A	966.3	longitude A.	5°55'W
latitude B.	23°35'S	mer. parts B	1447.4	longitude B.	24°30'W
d'lat	7°35'S	d.m.p.	481.1	d'long	18°35'W
=	455'S			=	1115'W

$$\text{tan course} = \frac{\text{d'long}}{\text{d.m.p.}} = \frac{1115}{481.1}$$

log. 1115	3.04728
log. 481.1	2.68224
log.tan co	0.36504

true course = S66°39.6'W = 246°39.6'

distance = d'lat x sec. course
= 455 x sec.66°39.6'

log. 455	2.65801
log.sec . 66°39!6	0.40210
log.distance	3.06011

Distance = 1148.5 miles

EXAMPLE 11

A ship left position lat 38°N., long 116°05'E and sails 1234 miles on a course of 322°T. Find her position at the end of the run.

dlat = distance x cos course.
= 1234 x cos 38°

log 1234	3.09132
log cos 38°	9.89653
log .dlat	2.98785
dlat =	972.4'N

using the dlat, it is possible to find the latitude of arrival position and the d.m.p.

lat. left	38°00.0'N	mer parts	2453.85
d'lat	16°12.4'N		
lat arrived	54°12.4'N	mer parts	3866.80
		d.m.p.	1412.95

dlong = d.m.p. x tan course
= 1412.95 x tan 38°

log 1412.95	3.15010
log tan 38°	9.89281
log dlong	3.04291
dlong =	1103.8

longitude left	116°05.0'E
d'long	18°23.8'W
longitude arrived	97°41.2'E

Answer: Final position lat. 54°12.4'N., long 97°41.2'E

MIDDLE LATITUDE SAILING

The figure shows the track of a vessel moving from position A to position B. The d'lat is equal to the arc of the meridian A'B or to the arc AB' but since the meridians are converging, the departure made good is less than the arc BB' and greater than the arc A'A.

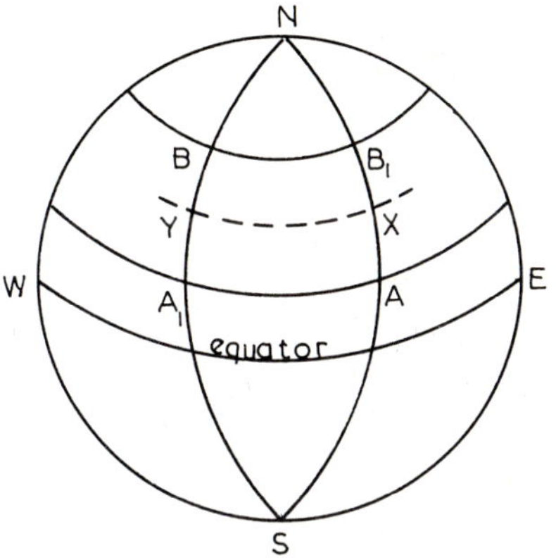

The distance made good in an East/West direction will be equal to the distance along some parallel XY which lies somewhere between the parallels through A and B.

The latitude of this parallel is the middle-latitude and from the formula previously established:

departure = d'long x cos.middle lat.

Where the d'lat is small and the latitudes involved are not high, the middle latitude may be taken as the mean latitude without appreciable error.

EXAMPLE 12

Find the departure between 30°N., 40°W. and 34°N., 36°W. using both mean and middle latitudes.

Using mean lat: departure = dlong.cos mean lat.
= 240 x cos 32°
= 203.5 miles

Using middle lat: mean lat. $32^\circ 00'.0$
correction $-\ 25'.8$
middle lat. $31^\circ 34'.2$

departure = dlong.cos mid.lat.
= 240 x cos.$31^\circ 34'.2$
= 204.5 miles

The correction to apply to the mean latitude to obtain the middle latitude was found in most navigational tables under the title of "Workman's correction". However, this correction was based on the assumption that the Earth is a perfect sphere. The table in Burton's and in the later editions of Norie's takes into account the spheroid shape of the Earth and consequently differs from Workman's correction which is now obsolete.

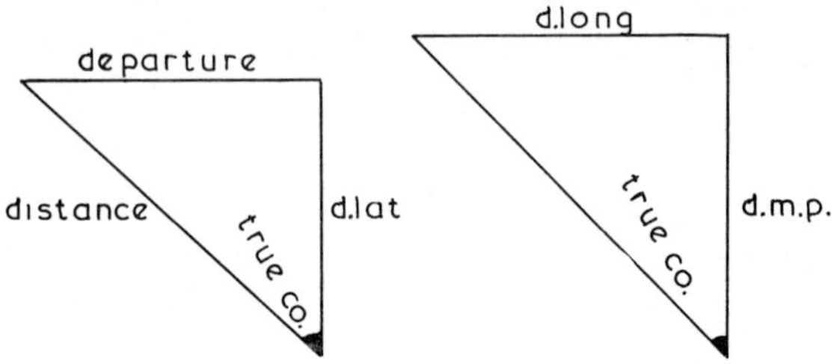

From the above diagram, it can be seen that the Mercator triangle is geometrically similar to the plane sailing triangle. Because of this similarity, the ratios between any corresponding sides must be a constant amount. Therefore:

$$\frac{\text{departure}}{\text{d'lat}} = \frac{\text{d'long}}{\text{d. m. p.}}$$

re-arranging $$\frac{\text{departure}}{\text{d'long}} = \frac{\text{d'lat}}{\text{d. m. p.}}$$

but $$\frac{\text{departure}}{\text{d'long}} = \text{cos. middle lat.}$$

therefore $$\text{cos. middle lat} = \frac{\text{d'lat}}{\text{d. m. p.}}$$

If the meridional parts for the sphere are used in the above formula, then middle latitude obtained is the same as that obtained from Workman's correction, whereas, if the meridional parts for the spheroid are used, then the middle latitude will be the same as that obtained from Burton's and other tables.

EXAMPLE 13

Find the correction to apply to the mean latitude 50° if the d'lat is 14°. The parallels of latitude are 43° and 57°

		(Spheroid)
latitude 43°	mer. parts	2847.1
latitude 57°	mer. parts	4163.0
d'lat 840'	d.m.p.	1315.9

$$\text{cosine middle lat.} = \frac{\text{d'lat}}{\text{d.m.p.}}$$

log 840	2.92428
log d.m.p. (spheroid)	3.11922
log cos.mid.lat.	9.80506

middle latitude	= 50° $19'.9$
mean latitude	= 50°
correction	$+19'.9$

From Burton's Tables the correction to apply is $20'$ +

In general, the middle latitude should be used when the distance exceeds 600 miles, which is the limit of the Traverse Table.

MISCELLANEOUS EXAMPLES ON THE SAILINGS

EXAMPLE 14

On a certain ship, the true course for the next four hours was 315°. After steaming for 60 miles it was found that the compass error of 9°W had been allowed the wrong way. Using the Traverse tables, calculate the distance of the ship from the estimated position.

True course required 315°
True course steered 297°

True course 315° x 60 miles. from T.T. d'lat 42'.4 N. dep 42'.4 W.
Course ste'd 297° x 60 miles, from T.T. d'lat 27'.2 N. dep 53'.5 W.

From A to B d'lat 15'.2 N. dep 11'.1 E

From the T.T. using the above d'lat and dep. Course N 36°E. dist. 18.8 miles

Answer: The vessel was 216° distant 18.8 miles from the E.P.

EXAMPLE 15

A vessel steams on a course of 300° for a distance of 400 miles and makes a d'long of 375'. Between which parallels did she sail?

From the T.T. course 300° distance 400 - d'lat 200' dep 346'.4

$\frac{\text{departure}}{\text{d'long}}$ = cos. middle lat.

	log 346.4	2.53958
	log 375	2.57403
log cos. middle lat.		9.96555
middle latitude		22° 31'.2N or S
correction, mean to middle lat.(Burton's)		+ 46'.0
mean latitude		23°17'.2 N. or S.
half d'lat		1°40'.0
1st parallel		21°37'.2N. or S.
2nd parallel		24°57'.2N. or S.

THE DAY'S WORK

This is simply the problem of finding the ship's course from noon to noon, average speed, set and drift. A typical, if somewhat complex, example follows:

EXAMPLE 16

At noon, 1st July, a vessel observed Cape Y (lat 43° 25'N., long 65° 30'W.) bearing 260° distant 14 miles. The log was set at 0 and the course was set to 325°. At midnight, log 178, course was altered to 021°. At noon, 2nd July, the log read 357. The current was assumed to set 348° at 2 knots throughout. Find the E.P. of the ship at noon, 2nd July, and the course and distance made good.

If the observed position of the ship at noon, 2nd July, had been lat 49° 46'N., long 66° 50'W. find the true set and drift experienced.

	D'lat		Dep.	
	N.	S.	E.	W.
Reversed bearing and distance - 080° x 14'	2.'4	-	13.'8	-
1st course and distance - 325° x 178'	145.'8	-	-	102.'1
2nd course and distance - 021° x 179'	167.'1	-	64.'1	-
			77.'9	102.'1W
				77.'9E
For D.R. pos. to find true set and drift	315.'3N			24.'2W
Set and drift allowed - 348° x 48'	47.'0			10.'0
For estimated position	362.'3N			34.'2W
Subtracting d'lat and dep. for bearing	2.'4			13.'8
For course and distance made good	359.'9N			48.'0W

TO FIND COURSE AND DISTANCE MADE GOOD

$$\frac{\text{departure}}{\text{d'lat}} = \text{tan. course}$$

distance = d'lat x sec. course

log 48.0	1.68124		log 359.9	2.55618
log 359.9	2.55618		log sec. course	0.00383
log tan	9.12506		log distance	2.56001

Course made good N7°36'W = 352°24'

distance made good 363.1 miles

TO FIND THE ESTIMATED POSITION

lat of Cape Y	43°25'0N	mer. parts	2881.3
d'lat	6°02.'3N		
lat. of E.P.	49°27.'3N	mer. parts	3406.1
		d.m.p.	524.8

d'long = d.m.p. x tan course

Long. of Cape Y	65°30.'0W	log d.m.p.	2.71999
d'long	1°10.'1W	log tan. course	9 12506
Long. of E.P.	66°40.'1W	log d'long	1.84505 = 70.'1

Estimated position at noon 2nd July - Latitude 49°27.'3N., Long 66°40.'1W.

TO FIND THE TRUE SET AND DRIFT

Cape Y lat.	43°25'.0N	mer. parts	2881.3	Cape Y long.	65°30'.0W
d'lat	5°15'.3N			d'long	1°52'.0W
D.R. lat	48°40'.3N	mer. parts	3334.6	D.R. long	67°22'.0W
		d.m.p.	453.3		

$\frac{\text{departure}}{\text{d'lat}}$ = tan. course

d'long = d.m.p. x tan. course

log	77'.9	1.89154
log	315'.3	2.49872
log tan co.		9.39282

log	453.3	2.65639
log tan.		9.39282
log d'long		2.04921 = 112'.0

2nd July, D.R. position noon	lat 48°40'.3N.	long 67°21'.9W
2nd July, Obs. position noon	lat 48°46'.0N.	long 66°50'.0W
	d'lat 1°05'.7N	d'long 31'.9W

mer. parts	3334.6
mer. parts	3434.9
d.m.p.	100.3

$\frac{\text{d'long}}{\text{d.m.p.}}$ = tan. set

log 31'.9	1.50379
log 100.3	2.00130
log tan. set	9.50249

Set = N 17°38'.6W = 342°21'.4

drift = d'lat x sec. set

log 65'.7	1.81757
log sec. set	0.02092
log drift	1.83849

drift = 68.9 miles

The traverse tables also offer an aid to solving that perpetual stumbling block, the four point bearing.

EXAMPLE **17**

Vessel steaming on a 326° compass course (dev. 6°W. var. 13°E.) observes a light four points on the starboard bow. After steaming for 1 hour at 14 knots the light was abeam. Wind NE. leeway estimated at 7°. current setting 245°T. at 3 knots. Find tne distance off the light at the beam bearing.

Compass course	326°
Error	7°E
True course	333°
Leeway	- 7°
Track	326°

From T.T.	Track 326° distance 14 miles	d'lat	11'.6N.	dep 7'.8 W.
	Set 245° drift 3 miles	d'lat	1'.3 S.	dep 2'.7 W.
	For course and dist. made good	d'lat	10'.3N.	dep 10'.5W

To find course and distance made good

$\frac{\text{departure}}{\text{d'lat}}$ = tan. course

log	10'.5	1.02119
log	10'.3	1.01284
log tan		0.00835

course made good = 314°27'T

dist. = d'lat sec. course

log 10'.3	1.01284
log sec. course	0.15472
log dist.	1.16756

distance made good = 14.71 miles

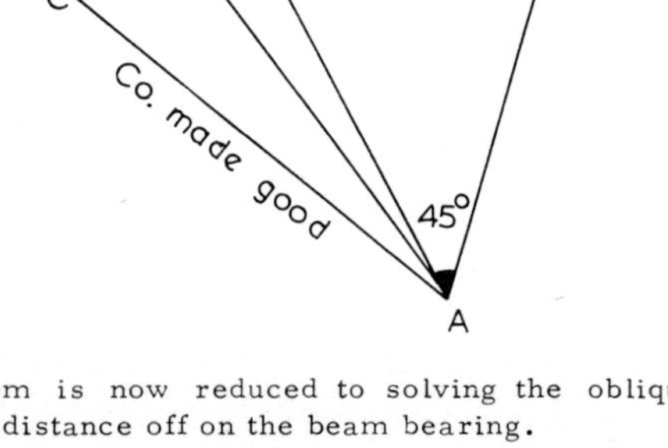

The problem is now reduced to solving the oblique angled triangle APC, where PC is the distance off on the beam bearing.

It should be noted that the course and distance made good is found by plane sailing formulae, rather than by Traverse Table, which restricts the accuracy required.

Triangle APC: $\dfrac{PC}{\sin 63^\circ 33} = \dfrac{AC}{\sin 45^\circ}$

and: $PC = AC \times \sin 63^\circ 33' \times \text{cosec } 45^\circ$

log AC 14.71	1.16756
log sin 63° 33'	9.95198
log cosec 45°	0.15051
log PC (distance off)	1.27006

Answer: Distance off on beam bearing 18.62 miles.

Alternative solution:

In diagram construct BD perp. to AF

and BE perp. to CP

hence: BD = EF, BE = DF, PF = AF,
angle BCE = 2° and angle DAB = 7°

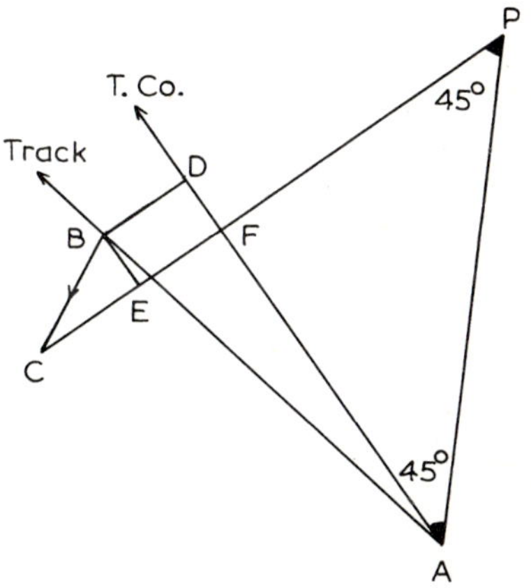

From Traverse Table:

Δ ABD AB:- 14' and DAB:- 7°
then AD = 13.9' and BD = 1.71'

Δ BCE BC:- 3' and BCE:- 2°
then CE = 2.99 and BE = 0.11'

then: PF = AD - DF
= AD - BE
= 13.9 - 0.11 = 13.79'

CF = CE + EF
= CE + BD
= 2.99 + 1.71 = 4.70'

therefore distance off = PF + CF = 13.79 + 4.70 = 18.49 miles

Answer: Distance off on beam bearing 18.49 miles.

IV
GREAT CIRCLE SAILINGS

THE GREAT CIRCLE SAILING

Any student who has some knowledge of geometry appreciates the fact that, the shortest distance between two points is a straight line. On the surface of a sphere, a straight line between two points is impossible. However, the shortest distance between two points on a sphere, will be the arc of a curve having the greatest radius. By definition, a great circle fulfills this condition, having the radius of the sphere itself. Therefore, the shortest distance between two points on the Earth's surface, will be the shortest arc of the great circle passing through the two points.

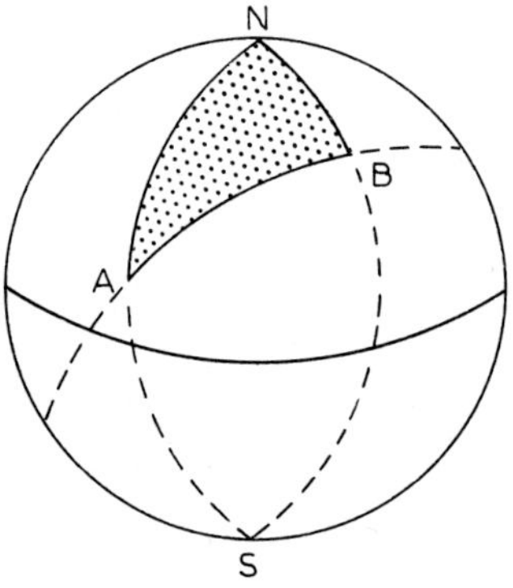

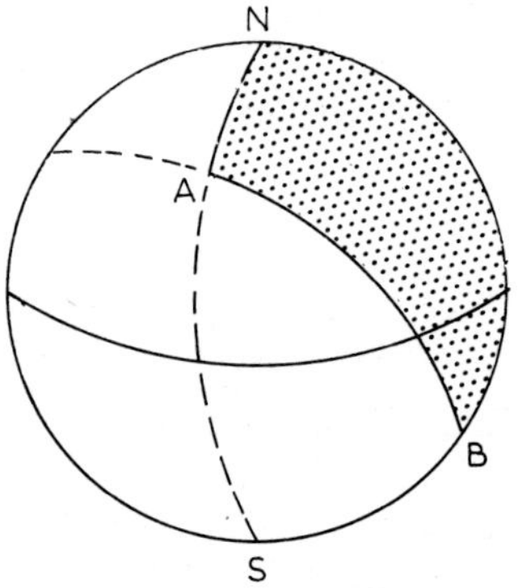

The figures show the arc of a great circle passing through points A and B. NA and NB are arcs of the meridians through A and B and are, therefore, arcs of great circles. The triangle NAB is a spherical triangle, and with three known quantities:

NA = co-lat of A
NB = co-lat of B
ANB = d'long between A and B

the solution of the triangle by haversine formula is relatively straightforward.

The closest approach of a great circle to the pole is known as the vertex of that great circle. At this point, the great circle cuts the meridian through the vertex at 90°.

Since the angle a great circle makes with the meridian is continually changing, a vessel, in following a great circle track, would have to continuously alter course. In practice, this is impossible, since a vessel must hold a steady course until a definite alteration is made. Consequently, a ship steams a series of rhumb lines between successive points on the great circle and makes an approximate great circle sailing.

The great circle track between two places will always pass nearer the pole than the rhumb line track (except on the equator); it follows that the ship may be taken into high latitudes where ice may be met. When this is likely, the track must be modified while still remaining the shortest track in the circumstances. This modified line is known as the 'composite track', and is formed by two different great circle arcs, joined by an arc of a parallel of 'limiting' latitude.

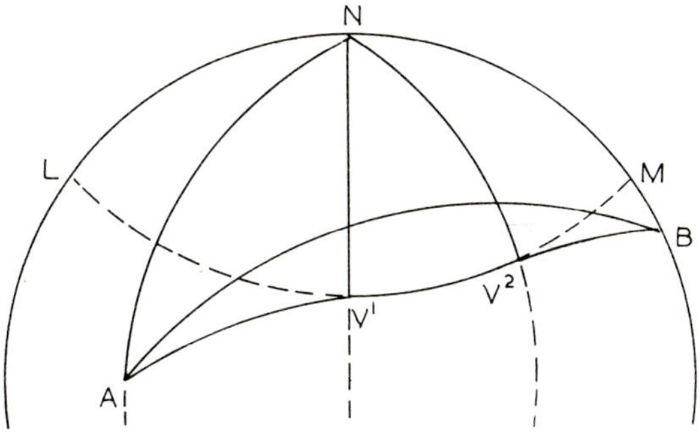

The figure shows such a composite track, with parallel L V^1V^2M being the limiting latitude.

The great circle arcs are tangential to the limiting parallel, so that V^1 and V^2 are the vertices of each great circle.

The longitude of the vertex differs by 90° from the point where the great circle crosses the equator. From this it follows that the course angle of the track at the equator is equal to the co-lat of the vertex.

To assist in finding the great circle track between two places, gnomonic charts are used, on which any straight line is a great circle.

The gnomonic chart is constructed by projecting the Earth's surface from the centre of the Earth on to a plane tangential to a convenient point. Since a great circle is formed by the intersection of a plane through the Earth's centre with the Earth's surface, and as one plane must always cut another in a straight line, all great circles will appear as straight lines.

Transference of a great circle track from a gnomonic chart to the normal navigational Mercator chart, is effected by simply noting the positions of convenient points, marking them on the Mercator chart, and then joining them by a smooth curve. In practice, the points are joined by a series of rhumb lines that the vessel will sail.

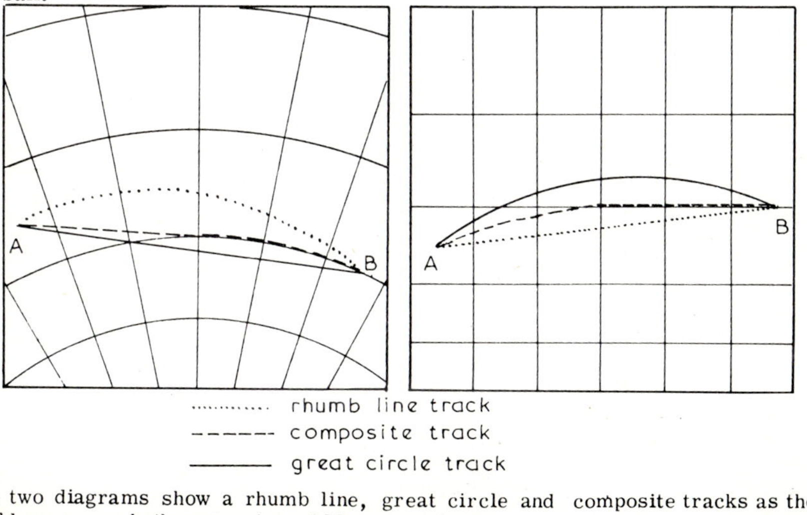

The two diagrams show a rhumb line, great circle and composite tracks as they would appear on both gnomonic and Mercator charts.

EXAMPLE **18**

Find the initial and final courses and the distance on the great circle track between Lat. 38°N., Long. 124°W. and Lat. 44°N., Long. 164°E. Find the position of the vertex and the latitudes in which the track crosses the meridians of 140°W., 160°W and 180°.

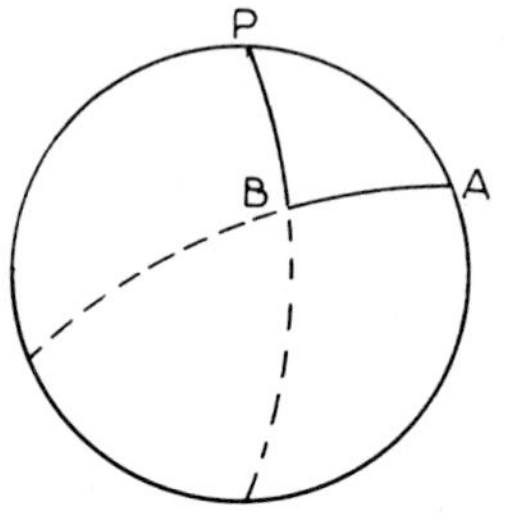

Lat 38°N.	PA $52^\circ 00'$	Long 124°W.
Lat 44°N.	PB $46^\circ 00'$	Long 164°E.
	PA ~ PB $6^\circ 00'$	P 72°W.

To find the distance

hav AB = hav P. sin PA. sin PB. + hav (PA ~ PB)

log hav P	9.53844
log sin PA	9.89653
log sin PB	9.85693
log hav	9.29190
nat hav	0.19584
hav (PA ~ PB)	0.00274
hav AB	0.19858

AB = distance = $52^\circ 55.6'$ = 3175.6 miles.

PA $52^\circ 00.0'$	PB $46^\circ 00.0'$
AB $52^\circ 55.6'$	AB $52^\circ 55.6'$
PA ~ AB $00^\circ 55.6'$	PB ~ AB $6^\circ 55.6'$

To find the initial course

$$\text{hav A} = \frac{\text{hav PB} - \text{hav (PA} \sim \text{AB)}}{\text{sin PA.sin AB}}$$

hav PB	0.15267
hav (PA ~ AB)	0.00007
hav	0.15260
log hav	9.18356
log cosec PA	0.10347
log cosec AB	0.09807
log hav A	9.38510

A = $59^\circ 01.9'$

Initial course = $300^\circ 58.1'$

To find the final course.

$$\text{hav B} = \frac{\text{hav PA} - \text{hav (PB} \sim \text{AB)}}{\text{sin PB.sin AB}}$$

hav PA	0.19217
hav (PB ~ AB)	0.00365
hav	0.18852
log hav	9.27535
log cosec PB	0.14307
log cosec AB	0.09807
log hav B	9.51649

B = $69^\circ 56.2'$

Final course = $249^\circ 56.2'$

To find the position of the vertex.

By Napiers Rules
sin PV = sin PA. sin A

log sin PA	9.89653	PV	$42^\circ 30.4'$
log sin A	9.93321		90°
log sin PV	9.82974	Lat.	$47^\circ 29.6'$N

cot P = cos PA.tan A

log cos PA	9.78934	P	$44^\circ 16.0'$
log tan A	0.22177	longA	$124^\circ 00.0'$W
log cot P	0.01111	long.	$168^\circ 16.0'$W.

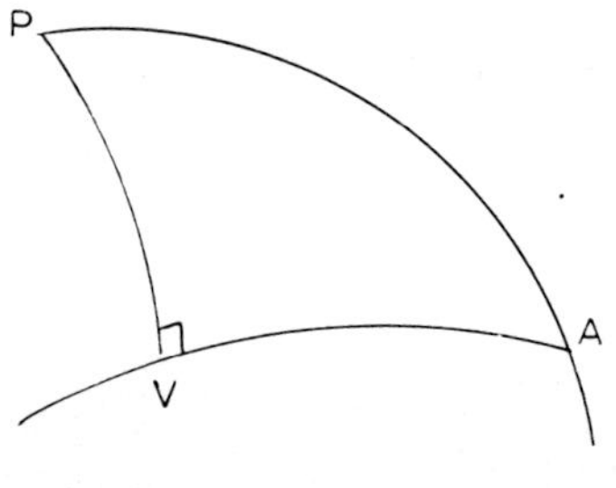

Owing to the small scale of the diagram, it is difficult to decide whether the vertex is between the departure and destination points - vertex 'in', or outside these two points - vertex 'out'. However, this difficulty can be overcome once the two courses have been calculated; if both angles are less than 90°, the vertex is 'in': if one is greater than 90°, the vertex is 'out' on the side of the greatest course angle. This will be appreciated if one bears in mind that on starting a great circle one steers towards the pole and on completing it one is steering away from it.

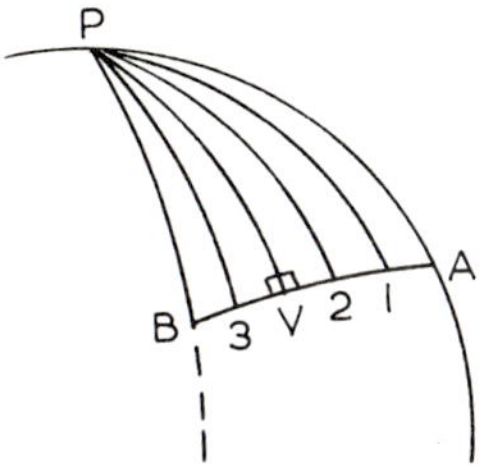

To find the latitudes of meridians.

In the diagram, P1, P2 and P3 represent the meridians of 140°W., 160°W. and 180°. Each meridian, together with the arc of the track and the meridian of the vertex, forms a right angled spherical triangle. The polar angle P is the d'long between the vertex and each meridian in turn.

In the triangle PV1 by Napier's rules

cot P1 = cos P.cot PV
and similarly for each meridian
cot P2 = cos P.cot PV
cot P3 = cos P.cot PV

long vertex	$168^{\circ}16'.0$W	long vertex	$168^{\circ}16'.0$W	long vertex	$168^{\circ}16'.0$W
long P1	$140^{\circ}00'.0$W	long P2	$160^{\circ}00'.0$W	long P3	$180^{\circ}00'.0$W
dlong P	$28^{\circ}16'.0$	dlong P	$8^{\circ}16'.0$	dlong P	$1^{\circ}44'.0$
log cot PV	0.03785	log cot PV	0.03785	log cot PV	0.03785
log cos P	9.94485	log cos P	9.99546	log cos P	9.99083
log cot P1	9.98270	log cot P2	0.03331	log cot P3	0.02868
P1	$46^{\circ}08'.5$	P2	$42^{\circ}48'.3$	P3	$43^{\circ}06'.6$
	90°		90°		90°
Latitude 1	$43^{\circ}51'.5$N	latitude 2	$47^{\circ}11'.7$N	latitude 3	$46^{\circ}53'.4$N.

EXAMPLE 19

Find the initial and final courses and the distance by composite track from A ($39^\circ 20'$S. $110^\circ 10'$E) to B ($44^\circ 30'$S., $46^\circ 20'$W), the limiting latitude being 62°S.

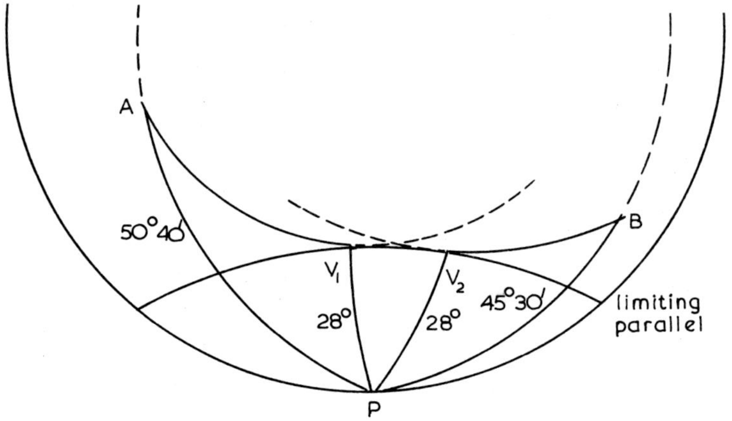

lat A $39^\circ 20'$S   lat B $44^\circ 30'$S
PA $50^\circ 40'$ PB $45^\circ 30'$

From right angled triangle PAV_1

Sin A = cosec PA. sin PV_1	cos P = cot PA. tan PV_1	cos AV_1 = cos PA. sec PV_1
log cosec PA 0.11156	log cot PA 9.91353	log cos PA 9.80197
log sin PV_1 9.67161	log tan PV_1 9.72567	log sec PV_1 0.05407
log sin A 9.78317	log cos P 9.63920	log cos AV_1 9.85604
= S$37^\circ 22'.3$E	P = $64^\circ 10'.3$	AV_1 = $44^\circ 07'.3$
	long A $110^\circ 10'.0$E	
Initial Course $142^\circ 37'.7$	long V_1 $45^\circ 59'.7$E	= 2647.3 miles

From right angled triangle PBV_2

sin B = cosec PB. sin PV_2	cos P = cot PB. tan PV_2	cos V_2B = cos PB. Sec PV_2
log cosec PB 0.14676	log cot PB 9.99242	log cos PB 9.84566
log sin PV_2 9.67161	log tan PV_2 9.72567	log sec PV_2 0.05407
log sin B 9.81837	log cos P 9.71809	log cos V_2B 9.89973
= N$41^\circ 09'.8$E	= $58^\circ 30'.0$	= $37^\circ 27'.3$
	long B $46^\circ 20'.0$W	
Final course $041^\circ 09'.8$	long V_2 $12^\circ 10'.0$E	= 2247.3 miles

The distance V_1V_2 is found from the parallel sailing formula:-

V_1V_2(dep) = dlong x cos. lat. (limiting parallel)

long V_1	$45^{\circ}59'.7$ W		log dlong	3.30746
long V_2	$12^{\circ}10'.0$ W		log cos lat	9.67161
dlong	$33^{\circ}49'.7$	= $2029'.7$	log dep	2.97907
				= 953.0 miles

Total distance = $AV_1 + V_1V_2 + V_2B$

= 2647.3 + 953.0 + 2247.3

= 5847.6 miles

The latitudes of successive points along the track may be calculated as shown in the previous example.

An alternative and more practical solution to the great circle sailing lies in the ABC Tables. To illustrate this method, the triangle used is taken from example 18.

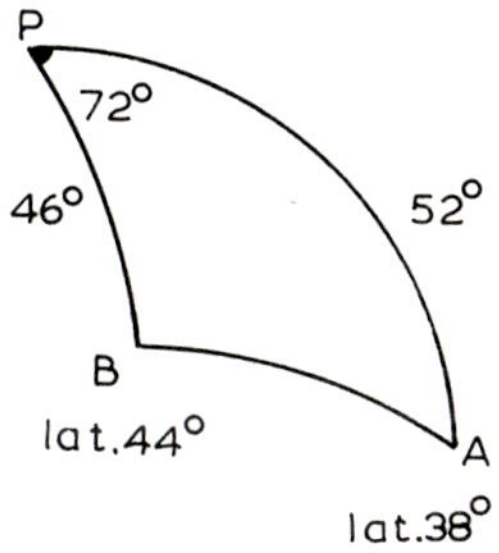

From Table A.

Using Latitude of departure point as Lat and dlong as Hour Angle A = 0.256 S

From Table B

Using Latitude of destination as Declination and dlong as Hour Angle B = 1.014 N

From A and B we obtain the value of C = 0.758 N.

Entering Table C with latitude of starting point and 0.758 N.

Azimuth = Initial Course = N $59^{\circ}.17$ W

= $300^{\circ}.83$

To find the final course the above procedure is reversed:-

From Table A.

Using lat of destination as lat. and dlong as Hour Angle

A = 0.310 S

From Table B.

Using lat of departure point as dec. and dlong as Hour Angle

B = 0.812 N

$A \sim B$ = C = 0.502 N

From Table C we obtain the azimuth and the reverse is the Final Course

= S $70^{\circ}.14$ W

= $250^{\circ}.14$

To find the distance (AB)

The position of the triangle is now adjusted so that the Hour Angle is the angle A. Angle P is the azimuth and side PA is the comp of the latitude.

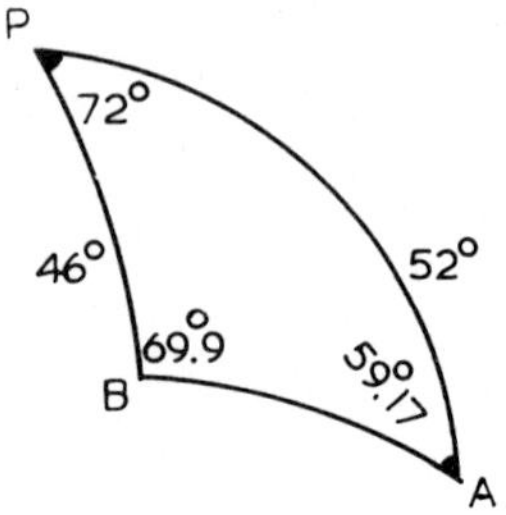

From Table C.

Using lat 38° and azimuth N 72°E   C = 0.412 N

From Table A.

Using lat 38° and Hour Angle 59°.17 A = 0.465 S

$B = A \pm C$

C = 0.412 N
A = 0.465 S
B = 0.877 N

From Table B.

Using B = 0.877 and Hour Angle 59°.17 → Dec = 36°.98
Remember that this dec. is the complement of the distance.

$$\therefore AB = 90^\circ - 36^\circ.98 = 53^\circ.02$$

$$\text{distance} = 53^\circ.02 \times 60 = 3181.2 \text{ miles}$$

This method, while not giving the accuracy of the longer method, is sufficient for practical purposes and affords a check when using the haversine method.

Other variations of the great circle sailing exist, particularly in examination problems.

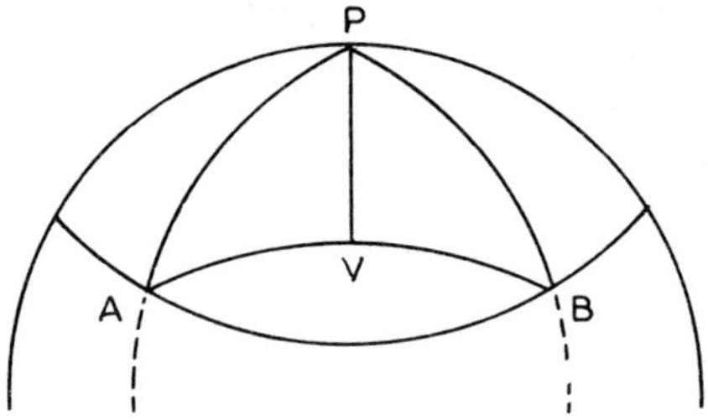

A typical one is where the starting point and the destination are in the same latitude. From the diagram it can be seen that the spherical triangle is isosceles. The solution is immediately reduced to one of right angled spherical trigonometry.

The vertex must lie in the middle of the track, consequently V is right angled in both triangles. The solution of one triangle also solves the remaining one, i.e:-

Initial course angle = final course angle
distance AB = 2.AV.

Another type often met with is where one position is given on the equator. This gives rise to the quadrantal triangle again solved by Napier's rules.

However, the position of the vertex is very easily found as can be seen from a diagram.

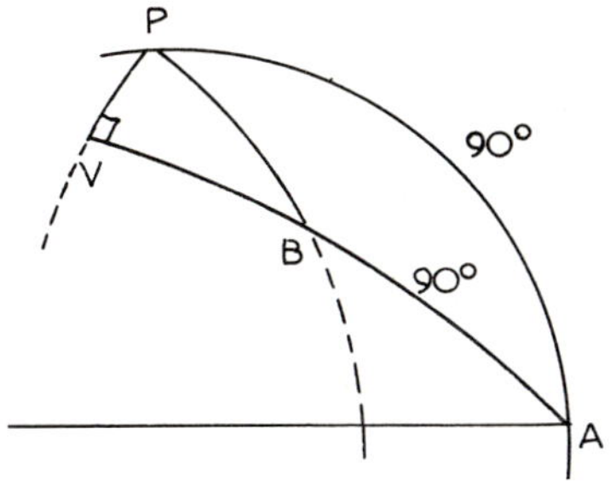

The co-lat of the vertex is equal to the course angle at the equatorial position.

The longitude of the vertex is 90° from the longitude of the position on the equator.

CONVERGENCY

When a radio station transmits a D. F. signal, the radio waves normally travel over the arc of a great circle between the station and the receiver. The bearing taken by the operator on the ship is, in effect, the "initial course" of the great circle track between ship and station.

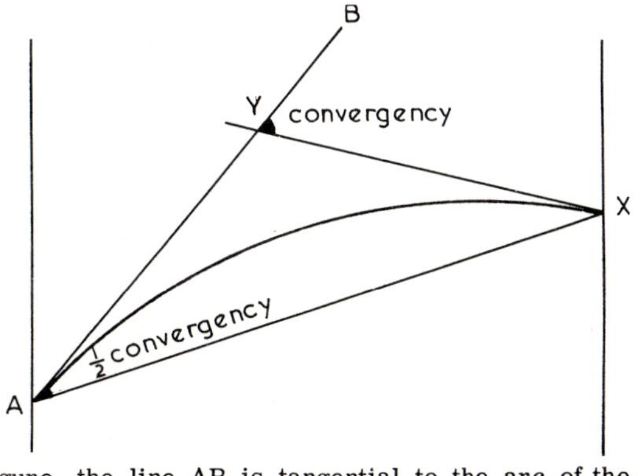

In the figure, the line AB is tangential to the arc of the great circle between A and X as it would appear on a Mercator chart. The direction AB is that shown on the dial of the D. F. set when receiving a signal from X. AX is the mercatorial bearing of X from A.

If the tangents to the great circle arc intersect at Y, then over relatively short distances (500 miles or less) the arc AX can be considered to form part of a true circle, and angles YAX and YXA are equal. The angle BYX is known as the convergency.

$$\begin{aligned}\text{Hence: convergency} &= \text{BYX}\\ &= \text{YAX} + \text{YXA}\\ &= 2 \times \text{YAX}\end{aligned}$$

Therefore the angle between the great circle bearing and the mercatorial bearing is equal to half the convergency.

The convergency may be found from the following formula:

convergency = dlong (in minutes) x sin. mean lat.

the result must be halved to obtain the half-convergency necessary to convert the D. F. bearing to a mercatorial bearing.

Since the great circle will always lie on the polar side of the mercatorial bearing then the half-convergency must always be applied towards the equator.

All navigational tables contain the correction to apply to D. F. bearings, i. e. half-convergency.

V
CELESTIAL SPHERE

THE CELESTIAL SPHERE

The orbit of the Earth around the Sun is in the form of an ellipse. When the Earth is at its farthest point from the Sun, it is said to be at 'aphelion' and the distance separating them is approximately 94.5 million miles. At the closest point the Earth is at 'perihelion' and the Sun is then about 91 million miles distant.

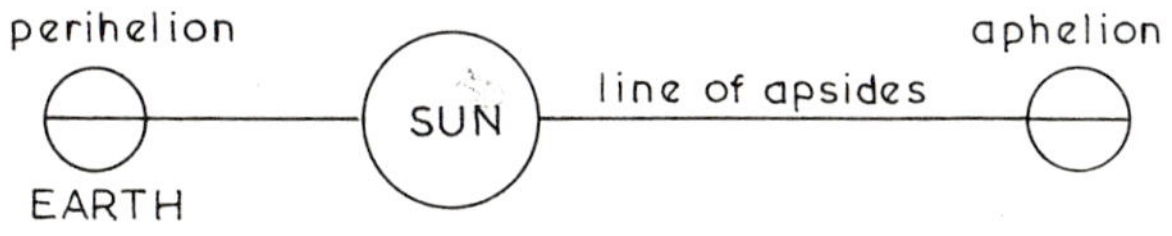

The line joining perihelion and aphelion is known as 'line of apsides' or 'aspe line'.

To observers on the surface of the Earth, the Sun and all the other heavenly bodies, irrespective of their actual distance, appear to revolve around the Earth as though situated on a sphere of immense radius with the Earth as its centre. This sphere is termed the 'celestial sphere'.

The stars, whose distances are measured in light years or in the astronomical unit 'parsec', maintain a fixed position on the celestial sphere, but the members of the solar system, which are relatively close to the Earth, appear to alter their respective positions from day to day.

If we consider the Earth orbitting around the Sun and if a given star is in the direction X, then as the Earth moves through positions 1, 2, 3, etc., the angle between the Sun and the star will gradually increase. The cycle will be completed in the time for the Earth to complete one orbit, i.e. one year. If this movement is transferred to the celestial sphere, the Sun will apparently move across the sphere returning to its original position in one year.

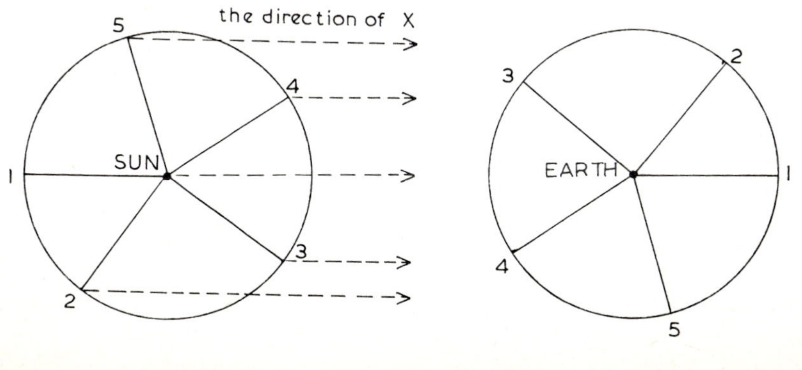

The following figure is for the purpose of illustrating the various points in the celestial sphere.

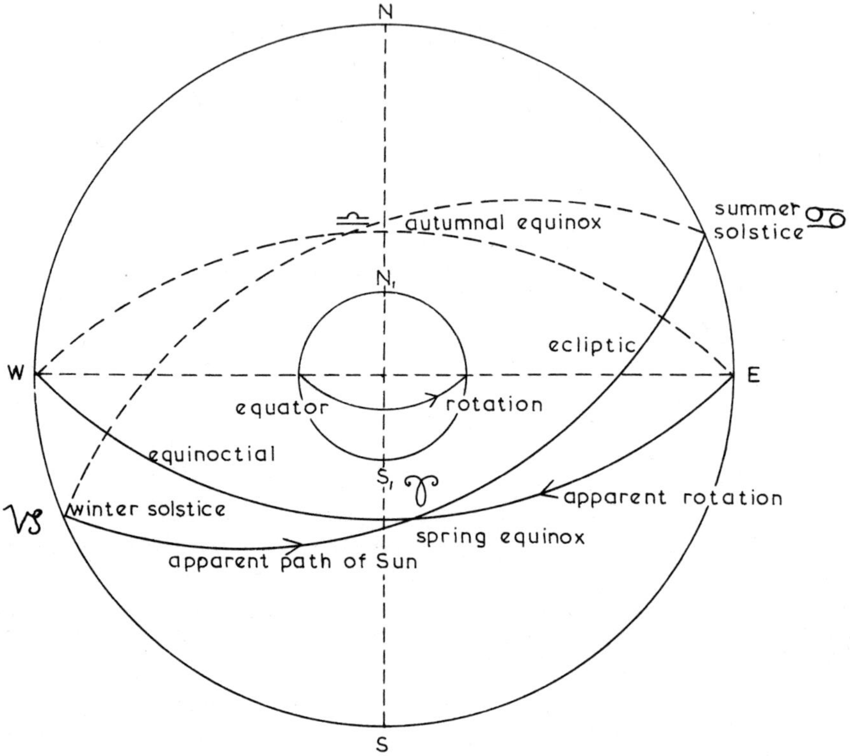

The celestial poles are the points N and S where the Earth's axis, if produced, would intersect the sphere.

The great circle WE in the plane of the Earth's equator is known as the Equinoctial or Celestial Equator.

From Spring Equinox (March 21st) to the Summer Solstice (June 22nd) is the season known as Spring for the Northern hemisphere. From Summer Solstice (June 22nd) to the Autumnal Equinox (Sept. 23rd) is the Summer season for the Northern hemisphere.

From Autumnal Equinox (Sept. 23rd) to the Winter Solstice (Dec. 22nd) is the Autumn season for the Northern hemisphere.

Finally from the Winter Solstice (Dec. 22nd) to the Spring Equinox (March 21st) is the Winter Season.

When the Sun is at the Equinoxes, i.e. its declination is 0° the duration of daylight and darkness is equal all over the world.

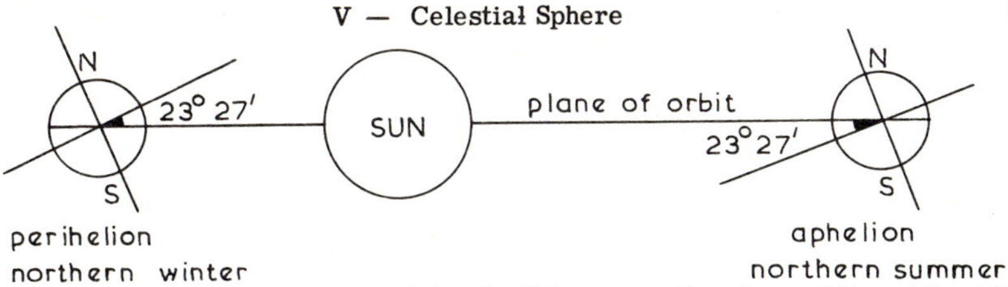

The figure shows the tilt of the Earth's axis to the plane of its orbit. This causes the apparent path of the Sun to form an angle with the equinoctial known as the Obliquity of the ecliptic. The ecliptic cuts the equinoctial in two places - the one through which the Sun passes on its passage north about the 21st March, is called the First Point of Aries (♈), or vernal equinox. The other, through which the Sun passes on its way south about the 23rd September, is called the First Point of Libra (♎), or autumnal equinox. The northernmost limit of its apparent path is the summer solstice and the southernmost limit is the winter solstice. These are also known as solstitial points.

The orbits of the planets are also inclined slightly to the plane of the Earth's equator, consequently their apparent paths are angled to the ecliptic. The points where they cross the ecliptic are known as nodes, where the planet crosses from south to north being known as the ascending node and from north to south as the descending node. The line joining nodes is called the nodal line.

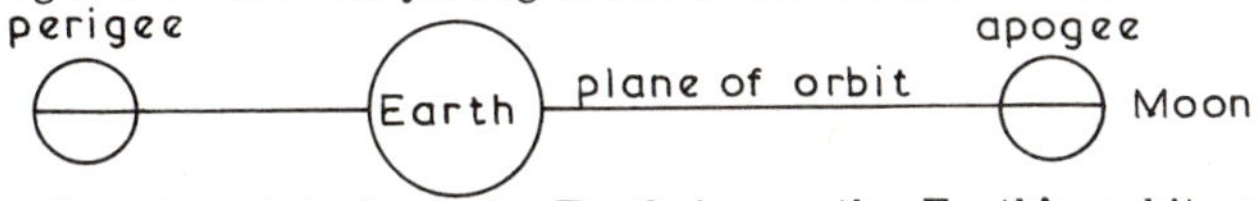

The Moon's orbit about the Earth is, as the Earth's orbit round the Sun, elliptical. The point of nearest approach, or perigee, is some 222,000 miles distant, while the farthest distance, apogee, is nearly 253,000 miles. The Moon in its apparent path around the celestial sphere, moves some 13° per day to the eastward against the background of the stars. The time taken to perform the complete revolution being $27\frac{1}{3}$ days.

THE PHASES OF THE MOON

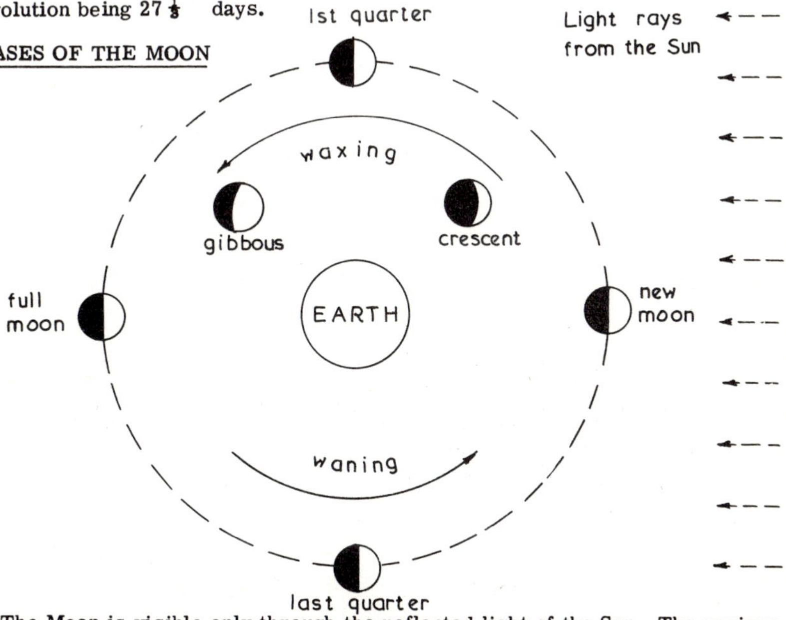

The Moon is visible only through the reflected light of the Sun. The various phases of the Moon are caused by its relative position to the Sun and the above figure is intended to illustrate this point.

ECLIPSES

Occasionally the Sun, Moon and Earth are in one straight line and when this happens an eclipse occurs. The eclipse will be either partial, total or annular and will only occur when the Moon is near or on the ecliptic, hence its name.

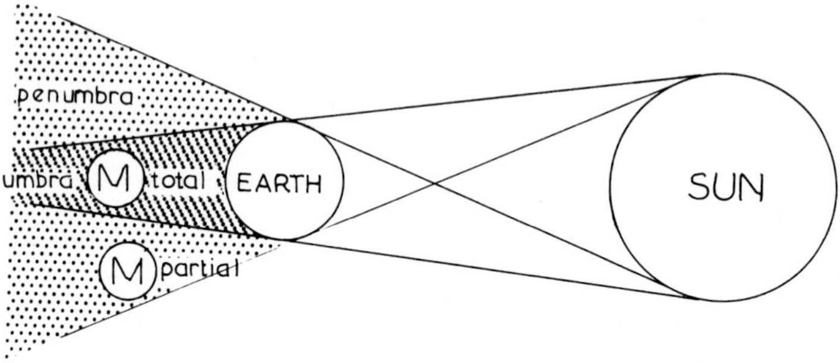

PARTIAL AND TOTAL ECLIPSE OF THE MOON

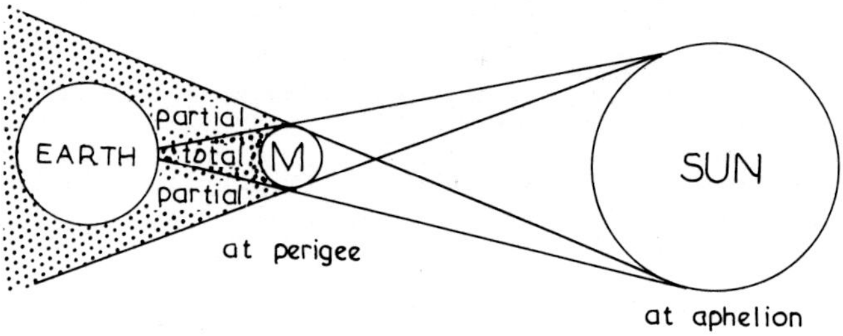

PARTIAL AND TOTAL ECLIPSE OF THE SUN

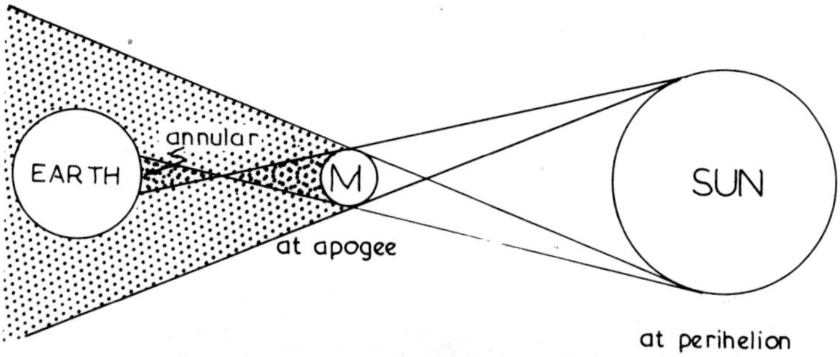

ANNULAR ECLIPSE OF THE SUN

During the Moon's passage of the celestial sphere, it may pass over a star or planet. This eclipse of the star or planet is known as an occultation.

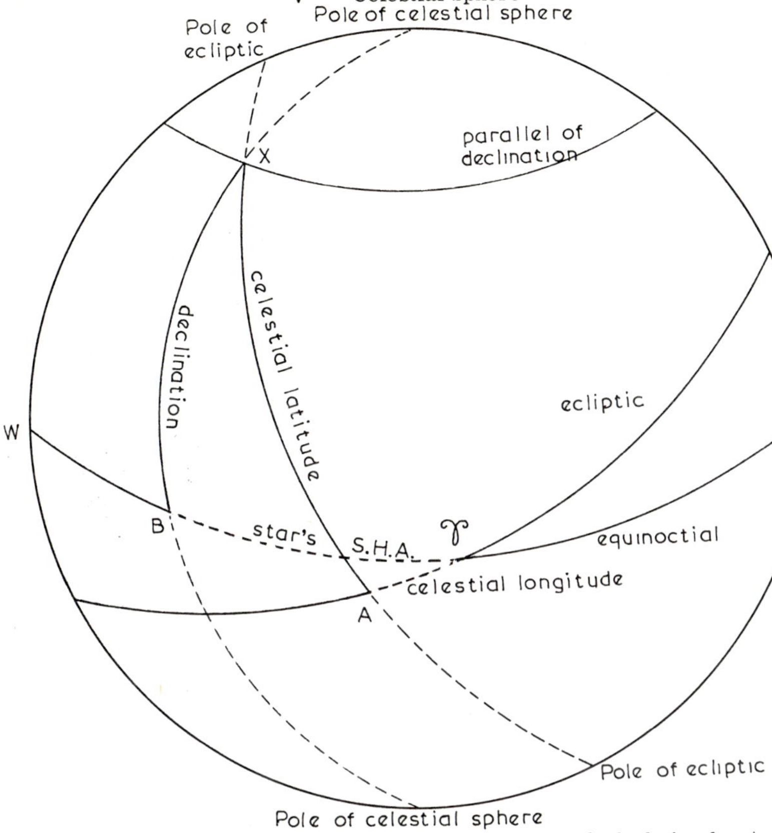

The sidereal hour angle (S. H. A.) of any heavenly body is denoted by the angle between the meridian through ♈ and the meridian through the body measured westwards from the former, and is expressed in units of angular measure.

When the angle between the meridian of ♈ and the meridian of the body is measured eastwards from ♈ , it is known as the Right Ascension (R. A.) of the body. This is more normally expressed in units of time.

Therefore R. A. = 360° - S. H. A.

For navigational purposes, R. A. has been replaced by S. H. A. but it is still used by astronomers for defining position on the celestial sphere.

Celestial longitude is measured along the ecliptic from ♈ to the meridian passing through the pole of the ecliptic (♈ A). It must never be confused with R. A. or S. H. A.

Declination, which corresponds to terrestrial latitude, is the angular distance of the body north or south of the equinoctial (BX). The declination of the stars changes very slowly, and together with S. H. A., is only tabulated in the Nautical Almanac at monthly intervals. The declination of the Sun changes quite rapidly, from $23\frac{1}{2}^{o}$N. to $23\frac{1}{2}^{o}$S. and back in twelve months. The declinations of the four navigational planets also vary between wide limits. These declinations are tabulated at hourly intervals, and interpolation is necessary for other times. To assist in interpolation, a quantity 'd' is given for the body concerned, and is equal to the mean hourly difference in the declination for that day.

A parallel of declination corresponds to a parallel of latitude and is a small circle on the celestial sphere parallel to the equinoctial.

Celestial latitude is measured north or south from the ecliptic along the meridian passing through the pole of the ecliptic (AX).

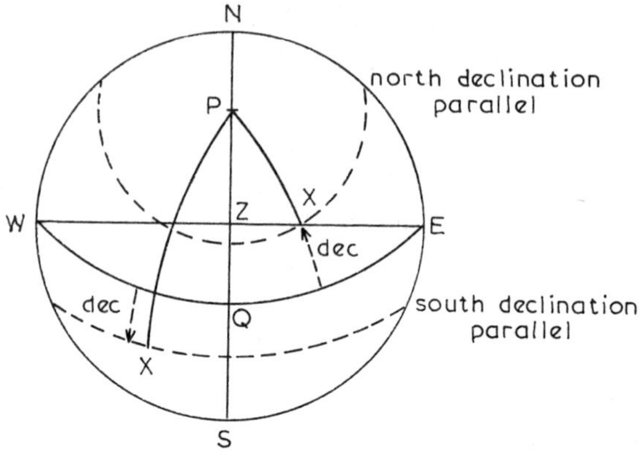

Polar distance is the angular distance of a body from the elevated pole, i.e. the pole above the observer's horizon. In the figure it is PX, with the observer assumed to be in north latitude.

When latitude and declination have the same names:- PX = 90° - dec.

When latitude and declination are of different names:- PX = 90° + dec.

The observer's zenith is the point on the celestial sphere vertically above the observer, Z. The declination of Z is then equal to the observer's latitude.

The celestial or rational horizon, is the great circle, every point of which is 90° from the observer's zenith, NESW.

The observer's meridian is the celestial meridian passing through the observer's zenith, NPZQS. Sometimes known as the principal vertical circle.

All great circles passing through the observer's zenith must be perpendicular to the rational horizon, and are known as vertical circles.

The prime vertical is the vertical circle passing through the east and west points, WZE.

The azimuth of a heavenly body is the angle at the zenith between the observer's meridian and the vertical circle through the body. It is measured east or west from the observer's meridian and named according to the elevated pole.

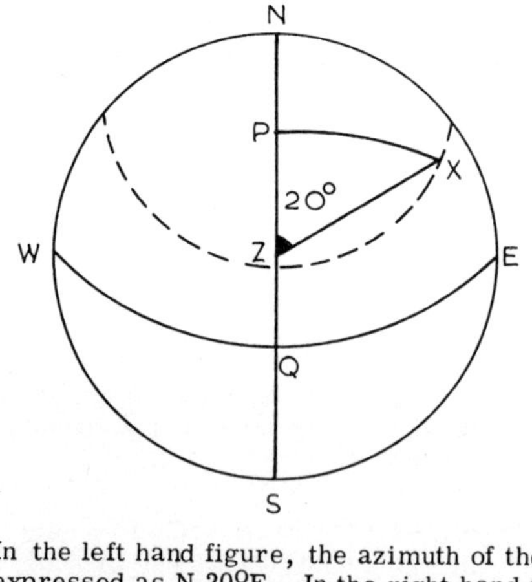

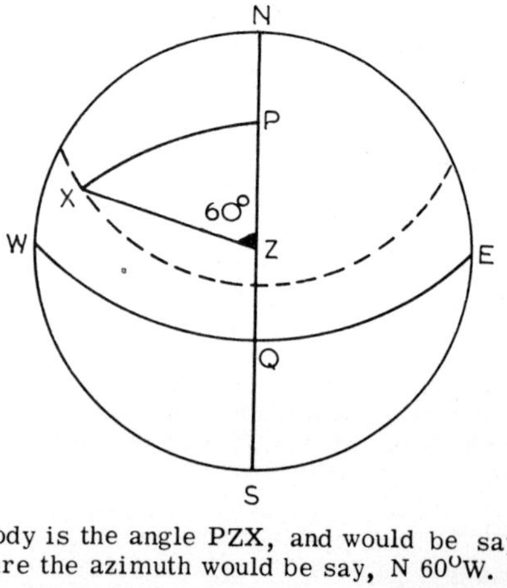

In the left hand figure, the azimuth of the body is the angle PZX, and would be say, expressed as N 20°E. In the right hand figure the azimuth would be say, N 60°W.

STELLAR SYSTEMS

To an observer on the Earth, the Sun and its satellites appear to move against a fixed background of stars. The largest of the Sun's satellites, or planets are Mercury, Venus, Earth, Mars, Jupiter, Saturn, Uranus, Neptune and Pluto. Mercury and Venus are known as inferior planets since their orbits are closer to the Sun than the Earth's. The remainder, with orbits outside that of the Earth, are known as superior planets.

Of the eight planets, only four - Venus, Jupiter, Mars and Saturn, in order of brightness - are used for navigational purposes.

Originally, stars were graded into six magnitudes, or measures of relative brightness, the sixth magnitude star being just visible to the naked eye. Later discoveries showed that a first magnitude star was about one hundred times brighter than a sixth magnitude star. Stars are now graded according to this definition, so that a second magnitude star is one hundred times as bright as a seventh magnitude star, and so on. Negative magnitudes become possible, since the star that is one hundred times as bright as a fourth magnitude star will have a magnitude of -1.

The intervening magnitudes between 1 and 6 can be found from a logarithmic scale:

Let a = the quantity of light received

then $\frac{a^6}{a} = \frac{100}{1}$ since the 1st mag. is 100 times brighter than the 6th mag.

therefore $a^5 = 100$
and $a = 2.51$ approximately

therefore a 1st. mag star is about $2\frac{1}{2}$ times as bright as a 2nd. mag. star, and so on.

For reference purposes, a table showing the magnitude of the brightest bodies is appended:

Name	Mag.	Name	Mag.	Name	Mag.
Sun	-26.7	Jupiter	-2.2	Capella	0.2
Moon (full)	-12.5	Mars	-0.2	Arcturus	0.2
Venus	- 3.4	Rigil. Kent.	0.1	Rigel	0.3
Sirius	- 1.6	Vega	0.1	Canopus	-0.9

Other stars of magnitudes brighter than 1 and which the navigator should be able to identify readily, are: Procyon, Achernar, Altair and Betelgeuse.

THE ZODIAC

The sun, moon and the principal planets have their apparent paths lying within a particular zone on either side of the equinoctial. This zone is known as the zodiac (Greek zodion - little animal). The zodiac was regarded with superstitious awe by the ancients; and the various movements of the sun, moon and planets through it still form the basis upon which astrologers claim to be able to predict the future and to cast horoscopes.

The diagram shows the signs of the zodiac together with symbols and Latin names.

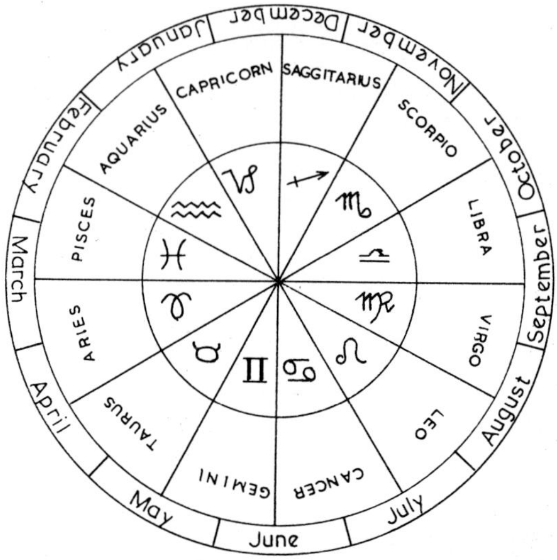

The signs of the zodiac were established about 2000 B. C. by the Babylonians. They thought that the equinoctial points were fixed, whereas in fact, they are slowly moving, making a complete circuit of the heavens in approximately 26,000 years.

As a result of the movement or precession, of the equinoxes, the sun is now a whole constellation behind the appropriate sign. For example, at the beginning of the spring the sun is said to enter the sign of Aries, which is now located in the constellation Pisces.

The following star charts of some of the principal constellations are drawn for the reader's guidance, their main purpose is to act as pointers to the brighter stars of navigational importance.

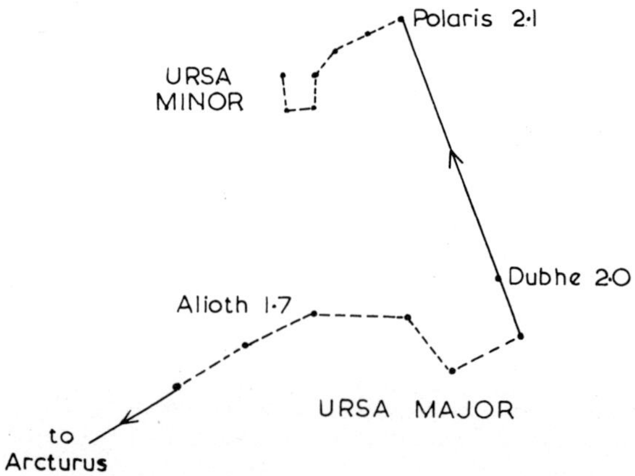

The chart showing the Plough (Ursa Major) shows the constellation below the pole, the observer facing north. By rotating the diagram through 180° the upper passage can be seen, with the observer still facing north.

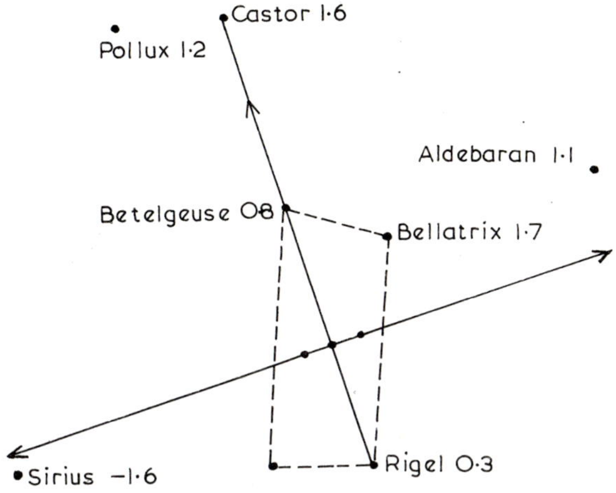

ORION

The constellation Orion is probably the most conspicuous in the sky. It offers pointers to Sirius, Aldebaran and the Heavenly Twins, (Castor and Pollux).

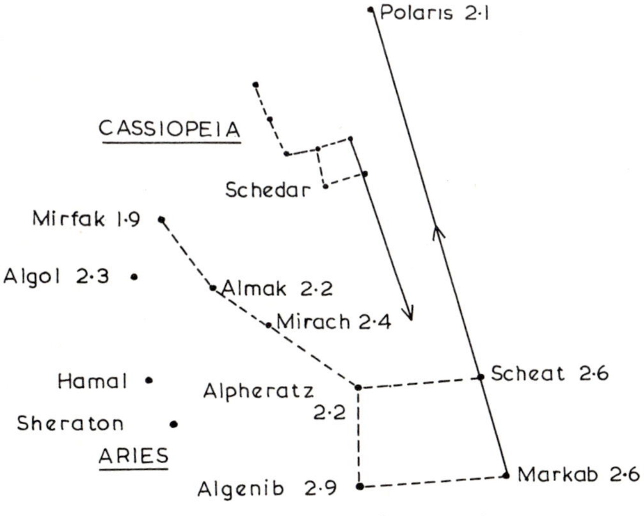

PEGASUS

The constellation Cassiopeia - sometimes known as the Chair - is found on the opposite side of the pole to The Plough and about the same distance away. While it does not contain stars of the first magnitude it helps the eye to pick up the constellation Pegasus.

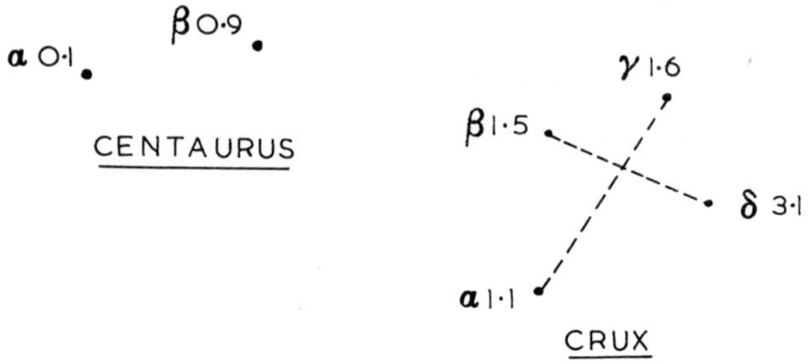

The constellation Southern Cross or Crux, while located in the Southern hemisphere, is too far from the celestial pole to be of any use in finding the latitude. Two bright stars in the constellation Centaurus direct the eye toward the Cross.

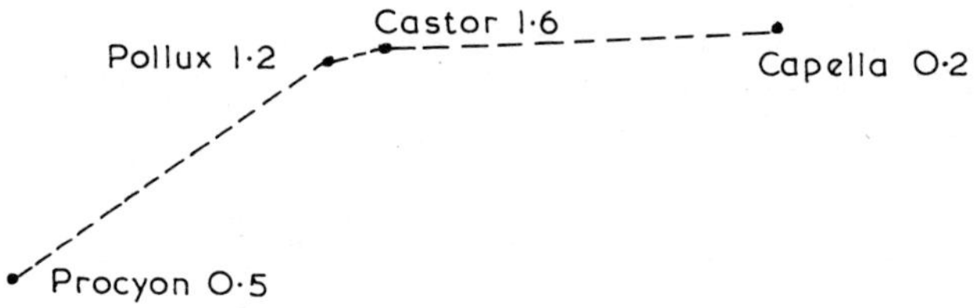

If the eye follows an arc clockwise from Procyon a number of navigational stars can be observed - Procyon, Pollux, Castor and Capella.

The star charts printed in the Abridged Nautical Almanac show all the stars that are likely to be used by the navigator, together with the relative positions of the major constellations to the Plough.

VI
TIME

CHRONOMETERS

In early days, the inability to determine the ship's longitude created many problems. Navigation consisted of fixes by bearings of shore objects, latitude sights and the use of the leadline.

Once out of sight of land and beyond the 100 fathom line, the early sailors relied on meridian altitudes and dead reckoning to determine their position - the result being many miles in error.

The governments of many countries offered substantial sums of money to anyone who would invent a method of determining longitude with a reasonable degree of accuracy. The largest was offered by the British Government:

£10,000	for a method accurate to 1°
£15,000	for a method accurate to 40'
£20,000	for a method accurate to 30'

Strange and varied was the response to these offers. One suggestion was to compare the magnetic variation at a given place with tabulated values for places of known longitude, but very large errors resulted from this. In 1714 a proposal was made to moor lightships at fixed intervals along the principal trade routes, which would fire star shells at regular intervals. At this time, it was thought that the maximum depth in the oceans was 300 fathoms!

However, it was recognised that, in order to solve the celestial triangle to find the longitude, it was necessary to have an accurate timekeeper. The winner of the £20,000 was John Harrison who made the first successful marine chronometer.

THE MODERN CHRONOMETER

The driving power is a mainspring of tempered steel wound in a spiral. The inner end is attached to the shaft or arbor, and the outer end attached to the containing drum. A thin chain coiled round the drum drives the fusee, the purpose of which is to maintain a constant drive to the chain of gears, known as the train. The train leads to the escapement, which gradually releases the power of the mainspring. The escapement is controlled by an oscillating balance wheel which beats once every half a second. Some balance wheels have a beat of one fifth of a second. The balance must be compensated for temperature changes, but with the discovery of some of the modern alloys, such as invar, which has a negligible coefficient of expansion, the compensation becomes less important.

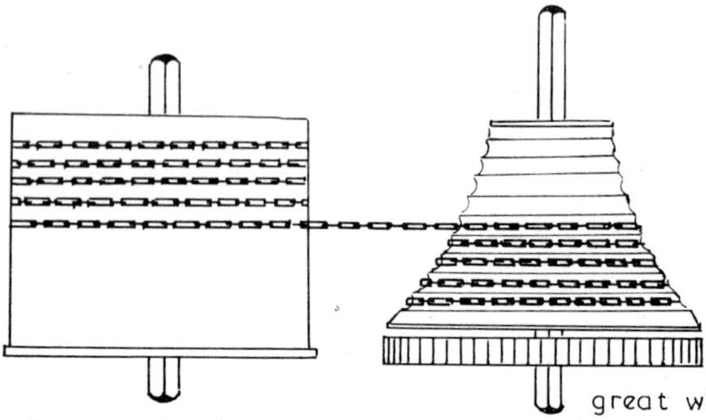

The train is made of brass, and the time of revolution of each wheel is determined by the number of teeth cut into its perimeter. The great wheel, which forms part of the fusee, drives the centre wheel which is attached to the hour hand. Another wheel drives the minute hand and another drives the second hand. There is a subsidiary pointer which indicates the tension in the mainspring and is known as the 'up and down' indicator. There is also a maintaining spring that keeps the chronometer going while it is being wound.

Chronometers are of the one day (30 hours), two day (54 hours), and eight day varieties.

CARE OF CHRONOMETERS.

Chronometers should be carefully stowed in their boxes, in a locker reserved for their exclusive use. The locker should have a glass top that is opened only for the purpose of winding the instrument. The locker should not be sited near single wire electric circuits or permanent magnets, such as compass corrector magnets. The chronometer itself must be slung in gimbals so that the dial is maintained in a horizontal position.

It is advisable to return chronometers to the manufacturer for cleaning and oiling at least every four years.

WINDING

Chronometers are wound by means of a 'tipsy' key. If the key is turned the wrong way, i.e. clockwise, the tipsy ratchet comes into operation to prevent damage to the chronometer.

The chronometer is carefully turned over by either gimbal, the dust cover is removed and then the chronometer is wound:-

A one day chronometer takes about 10 half turns per day.

A two day chronometer takes about $7\frac{1}{2}$ half turns per day.

An eight day chronometer takes about 4 half turns a day.

All chronometers on board should be wound at the same time, and by the same person each day. Any error that results from the movement and winding of the chronometer is then a constant factor in the daily rate.

Care should be exercised after winding in returning the chronometer to its original position by the same gimbal. It is possible to invert the chronometer by one gimbal and return it by the other, resulting in the dial appearing upside down.

STARTING

If the chronometer is accidentally allowed to stop, it should be turned by the inner gimbal until the dial is vertical and then rocked slightly, either forwards or backwards, by the outer gimbal. The inertia of the balance wheel will allow the escapement to unlock.

A chronometer that has stood idle for some time may suffer a change in its daily rate, and should, therefore, be re-rated.

On receipt of a new chronometer, the balance wheel will be wedged for safety during transit. The procedure is to unscrew the glass over the dial in an anti-clockwise direction; lock the gimbals, and then, with one hand on the dial, carefully turn the box over until the movement slides out. Should the movement be a tight fit, it can be started by inserting the key gently through the winding hole. The wedges should then be removed and the movement returned to its case. The chronometer is then wound and started.

SETTING THE HANDS

There is no valid objection, with modern chronometers, to re-setting the hands when the error is large, or on receipt of a new chronometer prior to starting it.

The following points should be noted:

1. The minute hand should be moved forward and never backward , as its mechanical advantage may damage the escapement.

2. The minute hand is turned by means of the winding key, never by means of finger pressure on the hand itself.

3. For a small alteration, it is not necessary to touch the hour hand, as it will automatically be carried round the correct amount by the minute hand. However, for large errors, such as six and a half hours, it is better to move the hour hand, gently and carefully, six hours, and then adjust the minute hand. The mechanical advantage of the hour hand is such that no strain is imposed on the escapement, and for the same reason, the minute hand will not follow its motion.

4. Never attempt to adjust the second hand.

ERRORS AND COMPARISON

In this modern day and age, W. T. time signals are an everyday event and afford an invaluable means of ascertaining the error of a given chronometer.

It is still possible, in some places, to check the chronometer by means of a visual time signal, but this method is rapidly falling into disuse.

Perhaps, one day, ships will carry a marine version of the quartz clock, or even a small version of the atomic clock whose accuracy is in the neighbourhood of a fraction of a second in several years!

When comparing two chronometers, it is usual to read one by sight and one by sound, - the tick being conventionally assumed to indicate the correct time. This is done by noting the instant at which a particular chronometer is, say, five seconds off the selected time and then counting the beats while observing the chronometer under comparison. It helps to distinguish the beat if the chronometer box is kept open, while the box of the chronometer being visually observed, is kept closed.

The following worked examples are typical of chronometer problems.

EXAMPLE 20

Chronometer A was 3m.30s. fast on chronometer B on 1st May at 1200 G.M.T. At 1200 G.M.T. on the 4th July, A was 00m.14s. fast on B. If the daily rate of A was 1.5s. gaining, find the daily rate of B.

1200 1st May to 1200 31st May	= 30 days
1200 31st May to 1200 30th June	= 30 days
1200 30th June to 1200 4th July	= 4 days
total time	= 64 days

Accumulated rate of A = 64 x 1.5 = 96 seconds gained.

chronometer A on 1st May was	3m. 30s. fast on B
chronometer A on 4th July was	0m. 14s. fast on B
therefore chronometer A has lost	3m. 16s. on B
accumulated rate of A	1m. 36s. gained
accumulated rate of B	4m. 52s. gained

therefore daily rate of B = $\frac{4m.52s.}{64} = \frac{292}{64} = \underline{4.56 \text{ seconds gaining}}$.

Note: If the chronometers are not compared at the same time G.M.T., then consideration must be taken of the fraction of a day involved when calculating the accumulated rate.

EXAMPLE 21

Chronometer A is 3m. 27s. fast on G.M.T., Chronometer B is 6m. 15s. slow on A, and chronometer C is 7m. 27s. fast on B. What is the error of B and C on G.M.T?

error of A =	3m. 27s. fast on G.M.T.
error of B =	6m. 15s. slow on A
therefore B is	2m. 48s. slow on G.M.T.
error of C =	7m. 27s. fast on B
therefore C is	4m. 39s. fast on G.M.T.

Answer: chronometer B is 2m. 48s. slow on G.M.T.
chronometer C is 4m. 39s. fast on G.M.T.

EXAMPLE 22

Star sights gave a vessel's position as lat. 36°05'N., long. 33°05'E. The error of the chronometer was assumed to be 4m. 15s. slow on G.M.T. After steering 204°T. for 11 miles, a lighthouse in position lat. 35°55'N., long. 32°40'E, bore 270°T. distant 6 miles. Find the true chronometer error.

	latitude	longitude	
star position	36°05'.0 N	33°05'.00 E	from T.T.204° by 11 miles
run	10'.0 S	5'.55 W	= dlat 10'.0 S
est. position	35°55'.0 N	32°59'.45 E	dep 4'.5 W = dlong 5'.55 W
lighthouse	35°55'.0 N	32°40'.00 E	from T.T.090° by 6 miles
rev. bearing	0.0	7'.32 E	= dlat 0'.0
obs. position	35°55'.0 N	32°47'.32 E	dep 6'.0 E
est. position	35°55'.0 N	32°59'.45 E	= dlong 7'.32 E
error in longitude	=	12'.13 W	

12'.13 arc = 48.4 seconds of time.

In the section on hour angles it is shown that:

Long W. = G.H.A. body = G.H.A.M.S. + 12 hours − equation of time. (for Sun)
= G.H.A.M.S. + 12 hours - S.H.A.M.S. + S.H.A.
(for star)

In these expressions, the equation of time and the S.H.A.s are constant for the observations and if the G.M.T. is too large the position line is displaced to the west, conversely, if the G.M.T. is too small the position line is displaced to the east.

In the problem, the position line has been displaced to the east, therefore the G.M.T. is too small.

Assumed error of chronometer	04m. 15s. slow	
additional error +	48.4s.	
correct error of chronometer	05m. 03.4s. slow	

EXAMPLE 23

Chronometer A gains 2.03s. daily, chronometer B loses 1.56s. daily. On 11th April A is 4m. 56s. slow on B. Compare the chronometers on 20th July, assuming no change in the daily rate.

(A further assumption must be made - that the two chronometers are compared at the same time G.M.T.)

11th April to 30th April	= 19 days
1st May to 31st May	= 31 days
1st June to 30th June	= 30 days
1st July to 20th July	= 20 days
total time	= 100 days

Accumulated rate of A = 100 x 2.03s. = 203s. gained
Accumulated rate of B = 100 x 1.56s. = 156s. lost

therefore A has gained 359s. on B.

error of A on B on 11th April = 4m. 56s. slow
total gain A = 5m. 59s.

error of A on B on 20th July = 1m. 03s. fast

THE HOUR ANGLE

The hour angle of a heavenly body is defined as the angle between the observer's meridian and the meridian through the body. It is measured westwards from the observer's meridian in either, units of arc, or units of time.

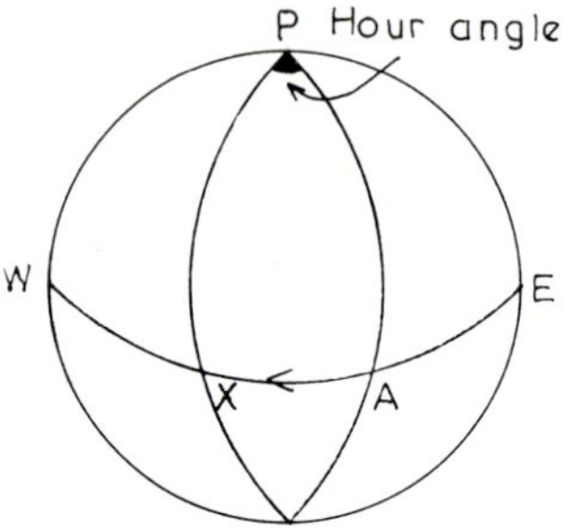

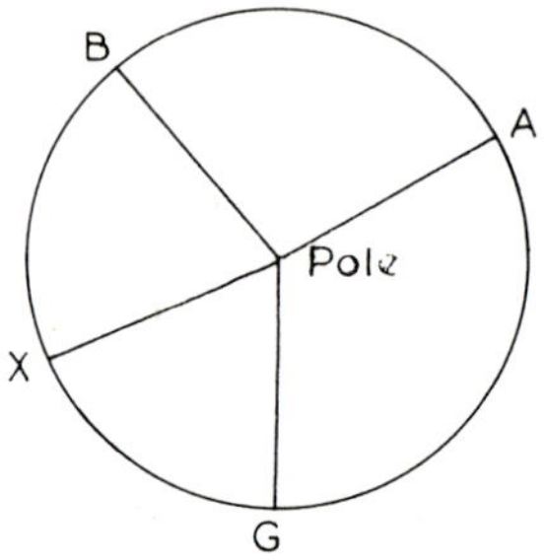

Since the angle at the pole is the hour angle, the angular length at the equinoctial AB also measures it.

When the observer is on the meridian of Greenwich, the hour angle is known as the Greenwich hour angle, or G. H. A. When the observer is on any other meridian, it is known as the local hour angle, or L. H. A.

In the case of the observer at A, (long. east of Greenwich):

G. H. A.	=	GX measured clockwise
Longitude	=	GA measured anti-clockwise
L. H. A.	=	AGX measured clockwise
	=	GX + GA
	=	G. H. A. + longitude

In the case of the observer at B, (long. west of Greenwich):

G. H. A.	=	GX measured clockwise
Longitude	=	GB measured clockwise
L. H. A.	=	GB + GX
	=	G. H. A. + (360° - longitude)
	=	G. H. A. - longitude, since the addition of 360° does not affect the actual angle.

Summarising:

To find the L. H. A. of a heavenly body

L. H. A. = G. H. A. + east longitude, subtracting 360° if necessary.
- west longitude, adding 360° if necessary.

When the L. H. A. of a heavenly body is less than 180°, the body will lie to the west of the observer's meridian, so that for L. H. A. s of 0° to 180°, the azimuth will be west, and conversely, for L. H. A. s greater than 180°, the azimuth will be east.

SOLAR TIME

The time taken by the Earth to complete one revolution about its own axis defines a "day". The time taken for the Sun to complete its apparent motion round the ecliptic defines a "year". These form the principal divisions of time. Any subdivision of these is quite arbitrary, and is simply for our convenience.

The interval between two successive transits of the Sun across the same meridian is an 'apparent solar day'. This is not an interval of fixed length because the Earth does not move along its orbit around the Sun at a constant speed. The time taken for the Earth to complete one revolution about its own axis is a fixed interval, but this will not be the length of a day as defined by the Sun.

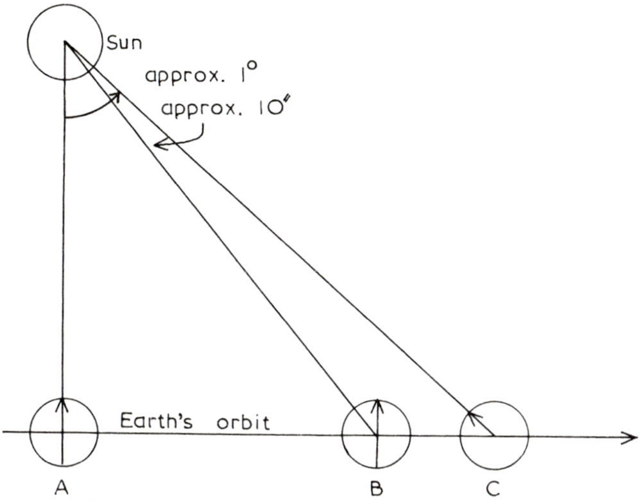

If the Sun is on the observer's meridian at A, then the Earth will have to complete something more than one revolution before the Sun returns to the observer's meridian, because the Earth will have moved along the path of its orbit. Unfortunately the distance moved by the Earth, A to C, is neither constant in length, nor is it described in constant time.

When the Sun is on the observer's meridian, it is said to be 'apparent noon and the Sun has attained its maximum altitude if the observer is stationary.

It would seem that the local hour angle of the Sun would form a convenient measure of time, but for practical purposes, it is better that the day should start at midnight, therefore:-

Local Hour Angle True Sun (L.H.A.T.S.) $\pm$ 12 hours = apparent solar time

Since the apparent solar day is not a constant interval, it has led to the introduction of a 'mean sun'. It is an imaginary body that is assumed to move along the celestial equator, or equinoctial, at a uniform speed, and to complete one revolution in the time taken for the true Sun to complete one revolution in the ecliptic.

The mean solar time is the local hour angle of the mean Sun (L.H.A.M.S.) $\pm$ 12 hours.

The mean solar day is the interval between successive transits of the mean Sun across the same meridian. This is the basis of everyday units of time.

The mean time kept at any place (Local Mean Time) is the hour angle of the Mean Sun measured from the meridian of that place.

L.M.T. = L.H.A.M.S. $\pm$ 12 hours

The local mean time kept on the meridian of Greenwich is known as Greenwich Mean Time.

G.M.T. = G.H.A.M.S. $\pm$ 12 hours.

A particular problem that arises with the use of the imaginary Mean Sun is that, an observation of the True Sun is timed by means of a chronometer that keeps mean solar time. There must, then, be some means of equating apparent solar time with mean solar time. The connection is known as the Equation of Time, and is defined as the excess of mean time over apparent time:

Equation of time = L. H. A. M. S. - L. H. A. T. S.

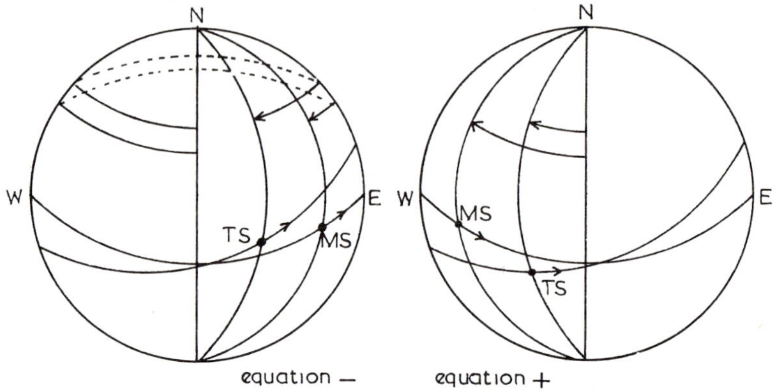

The figures show that at certain times L. H. A. T. S. will be either greater or less than L. H. A. M. S. Therefore the equation of time will be either positive or negative. The equation of time is zero on, or about, 15th April, 14th June, 1st September and 24th December. Its extreme values are +14 and -16 minutes.

It can also be defined as the excess of the Greenwich hour angle Mean Sun over the Greenwich hour angle True Sun.

Therefore:	G. H. A. M. S. − G. H. A. T. S. = equation of time
or:	G. H. A. T. S. = G. H. A. M. S. − equation of time
and:	G. H. A. T. S. = G. M. T. ± 12 hours − equation of time
therefore:	L. H. A. T. S. = G. M. T. ± 12 hours [+ long E. / − long W.] − equation of time.

The variations of the equation of time are due to the obliquity of the ecliptic and the eccentricity of the Earth's orbit around the Sun.

The figure below illustrates these two components, which, when combined, produce the equation of time.

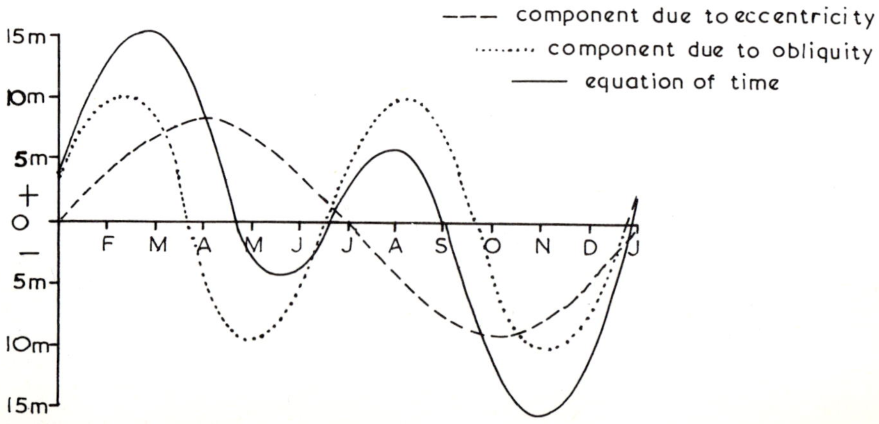

The equation of time can be found on the daily pages of the Abridged Nautical Almanac.

The problem of the L. H. A. T. S. is very much simplified by simply looking up the G. H. A. T. S. in the daily pages of the Almanac, where it is printed as an angular measure, and then applying the longitude as an angular measure.

Since all places on the Earth are identified by reference to the Greenwich meridian, i. e. longitude, then this provides a connection between L. M. T. and G. M. T. The Mean Sun covers an angular measure Westwards of 360° in 24 hours or 15° in 1 hour. Therefore a place that is 15° west of the Greenwich meridian will be 1 hour behind G. M. T. Similarly a place 15° to the east of Greenwich will be one hour ahead of G. M. T. The procedure for converting L. M. T. to G. M. T. is as follows:

Express the longitude of the place as units of time.
Subtract this from the L. M. T. if in east longitude.
Add if in west longitude.

A very popular aide-memoire:

Longitude west, Greenwich time best.
Longitude east, Greenwich time least.

In converting degrees to hours and minutes, the procedure is as follows:

Divide the degrees by 15 - this gives the number of hours.
Express the remainder as minutes of arc and divide by 15 - this gives the minutes of time.

Express the final remainder as seconds of arc and divide by 15 to obtain seconds of time.

EXAMPLE **24**

Convert the following longitudes to units of time: $100^{o}24'30''$, $78^{o}34'30''$.

$15)100^{o}$ 24' 30'' (6 hours
90
10° = 600'
15)624' 30'' (41 minutes
615
9' = 540''
15)570 (38 seconds
570

$100^{o}24'30''$ = 6hrs.41mins.38secs.

$15)78^{o}$ 34' 30'' (5 hours
75
3° = 180'
15)214' 30'' (14 minutes
210
4' = 240''
15)270'' (18 seconds
270''

$78^{o}34'30''$ = 5hrs.14mins.18secs.

The conversion of time to longitude is the reverse of the above procedure, i.e. multiplying by 15.

EXAMPLE 25

The L.M.T. in longitude 35° 30'E. is 15h.12m.24s. Find the G.M.T.

	L.M.T. 15h.12m.24s.
long expressed as time	2h.22m.00s.
	G.M.T. 12h.50m.24s.

EXAMPLE 26

On the 3rd April the G.M.T. was 02h.56m.16s. in longitude 74°26'W. Find the L.M.T.

G.M.T.	3d.02h.56m.16s.
long expressed as time	4h.57m.44s.
L.M.T.	2d.21h.58m.32s. (note change of date)

EXAMPLE 27

On the 5th December the G.M.T. at a particular place was 23h.56m.45s. and the L.M.T. for the same date was 15h.30m.15s. Find the longitude.

G.M.T.	5d.23h.56m.45s.
L.M.T.	5d.15h.30m.15s.
long expressed as time	08h.26m.30s.

and since G.M.T. is 'best' the longitude must be west

Longitude 126°37'30"W.

It would be most impracticable for each place to keep the local time of its meridian, or for all places to keep the same time. The compromise is for a group of places in the same locality - often a whole country - to keep what is known as the Standard Time of that area. The standard times kept by various countries are listed in the Nautical Almanac and in the Admiralty List of Radio Signals, Volume II.

Zone time is an extension of standard time and is kept by the majority of ships at sea. The Earth is divided into zones of 15° longitude and situated so that the central meridian of each zone is an exact number of hours distant from the Greenwich meridian. Zone time differs from G. M. T. by an exact number of hours and is fast or slow depending whether east or west of Greenwich, respectively. The zone can be found by adding 7° 30' to the longitude and then dividing by 15°, the resulting whole number is the zone.

EXAMPLE 28

In which zones are longitudes 130° E and 82° 10' W situated ?

	130° 00' E			82° 10' W	
	7° 30'			7° 30'	
15°)	137° 30'	(9	15°)	89° 40'	(5
	135°			75°	
	2° 30'			14° 40'	

Zone is -9   Zone is +5

NOTE. The sign given to the zone indicates the way in which the zone number is applied to zone time to obtain G.M.T.

SIDEREAL TIME

With solar time, the governing body was the Sun, and the day was defined as the interval between successive transits of the Sun. If the body is a star, the interval between successive transits is known as a sidereal day. In practice, the First Point of Aries is used instead of an actual star, and the sidereal day is further defined as the interval between two successive transits of the First Point of Aries across the same meridian.

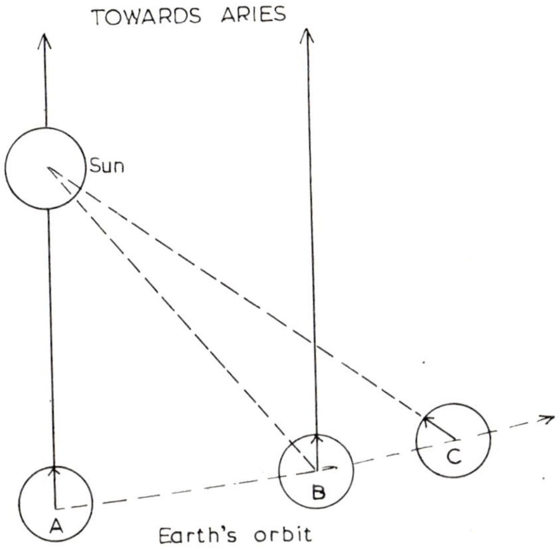

When the Earth has rotated through 360° and has reached B along its orbit Aries is again on the meridian. The Sun will not be on the meridian until the Earth reaches position C, i.e. the Earth rotates through about 361°. The sidereal day is therefore, nearly 4 minutes shorter than the solar day. The only importance of sidereal time to the navigator is in finding a given star's hour angle.

The other definitions that must be borne in mind when calculating a star's hour angle are:

Local hour angle First Point of Aries :- the angular distance of Aries from the observer's meridian, and similarly, the Greenwich hour angle of the First Point of Aries is the angular distance of Aries from the meridian of Greenwich, measured westwards. Sidereal hour angle (S. H. A.) has already been defined on page 41.

The G. H. A. Aries is tabulated against G. M. T. in the Abridged Nautical Almanac on the daily pages together with a quantity that remains practically constant - the S. H. A. of some 40 fixed stars. A further list of S. H. A. s is given at the back of the N. A.

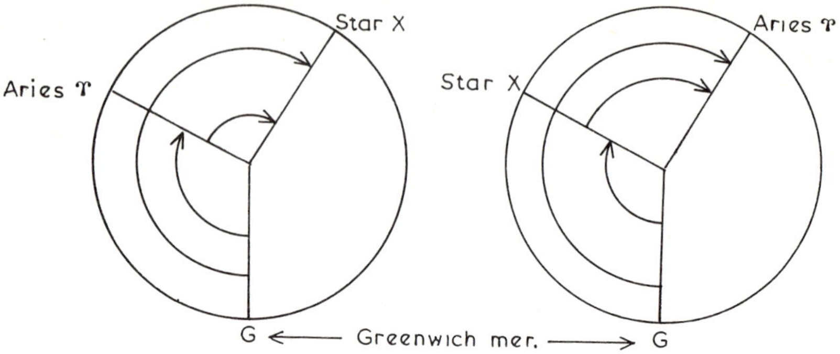

In the left hand figure:

GX = G♈ + ♈X (all measured westerly)

or: G.H.A. star = G.H.A. Aries + S.H.A. star

In the right hand figure:

GX = G♈ − ♈X

or G.H.A. star = G.H.A. Aries − (360° − S.H.A. star)
= G.H.A. Aries − 360° + S.H.A. star

and since the 360° does not affect the actual angular measure, then in all cases:

G.H.A. star = G.H.A. Aries + S.H.A. star

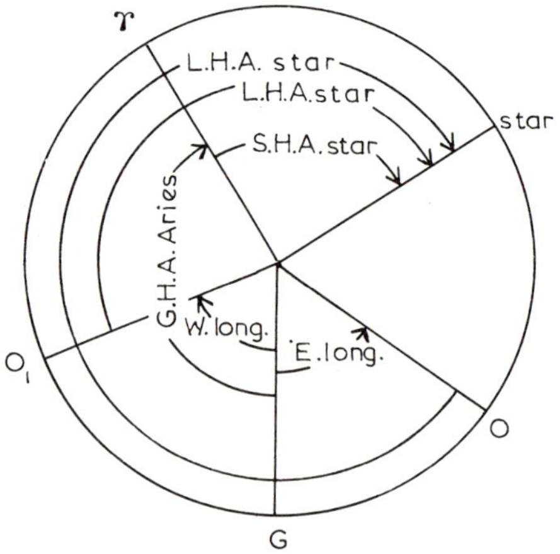

It must follow that the local hour angle of a star will be the Greenwich hour angle with the correct application of the longitude. The diagram shows an observer in east longitude (0) and one in west longitude (0_1)

L.H.A. star = G.H.A. star − W. longitude
= G.H.A. Aries + S.H.A. star − W. longitude (for 0_1)

or: L.H.A. star = G.H.A. Aries + S.H.A. star + E. longitude (for 0)

An adaptation of the expression can be used when it is required to forecast the passage of a particular star across the meridian for a meridian altitude.

When a star is on the observer's meridian its hour angle is either 000° for upper transit or 180° for lower transit.

Considering an observer in west longitude:

L. H. A. star = G. H. A. Aries + S. H. A. star − W. longitude

therefore 000° = G. H. A. Aries + S. H. A. star − W. longitude

or W. longitude = G. H. A. Aries + S. H. A. star

Since the S. H. A. star can be taken from the Almanac, then

W. longitude − S. H. A. star = G. H. A. Aries

By reference to the Almanac the G. M. T. for that particular G. H. A. Aries can be found and hence the G. M. T. of the star's transit.

If the observer is in east longitude, then the east longitude must be converted to west longitude, a relatively simple matter. For the lower transit of the star, 180° must be added to the west longitude.

EXAMPLE 29

Find the G.M.T. of upper transit of Achernar on 10th June 1958 in lat. $8^{\circ}40'$N. long. $27^{\circ}30'$W.

W. long =	G.H.A. star	$027^{\circ}30'.0$	
	S.H.A. star	$335^{\circ}57'.9$	
	G.H.A. ♈	$51^{\circ}32'.1$	
from Nautical Almanac		$48^{\circ}15'.2$	- 10d.10h.00m.00s.
	increment	$3^{\circ}16'.9$	- - - 13m.06s.
	G.M.T. of upper meridian passage		10d.10h.13m.06s.

EXAMPLE 30

Find the L.M.T. of lower transit of Kochab on 13th November 1958 in lat. $49^{\circ}20'$N. long. $127^{\circ}15'$W.

(Note: L.H.A. star on lower transit = 180°)

W. long + 180° =	G.H.A. star =	$307^{\circ}15'.0$	
	S.H.A. star =	$137^{\circ}19'.5$	
	G.H.A. ♈	$169^{\circ}55'.5$	
	from Nautical Almanac	$157^{\circ}52'.6$	- 14d.07h.00m.00s.
	increment	$12^{\circ}02'.9$	- - - 48m.12s.
	G.M.T. of lower meridian passage		14d.07h.48m.12s.
	longitude		8h.29m.00s.
	L.M.T. of lower meridian passage		13d.23h.19m.12s.

EXAMPLE **31**

Find the zone time of the upper transit of Regulus on 3rd March 1958 in lat.$12^\circ 15'$N. long. $100^\circ 27'$E.

W.long = 360° - E.long =	G.H.A. star =	$259^\circ 33'.0$	
	S.H.A. star =	$208^\circ 27'.4$	
	G.H.A. ♈	$51^\circ 05'.6$	
from Nautical Almanac		$41^\circ 54'.4$	4d.16h.00m.00s.
	increment	$9^\circ 11'.2$	- - 36m.39s.
G.M.T. of upper meridian passage			4d.16h.36m.39s.
long.$100^\circ 27'$E = zone - 7			+ 7h.00m.00s.
zone time of upper meridian passage			4d.23h.36m.39s.

The lunar day is different from either the solar or the sidereal day since the Moon itself is orbitting around the Earth in the same direction as the Earth's spin. Consequently the lunar day is slightly longer than the solar day.

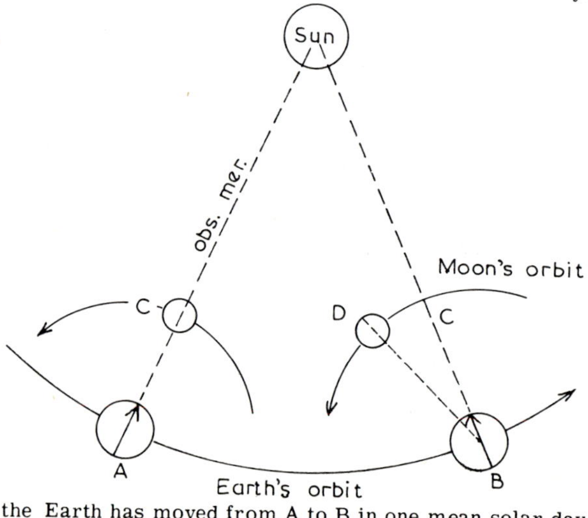

In the figure, the Earth has moved from A to B in one mean solar day and the moon will have moved from C to D. Therefore the Earth must rotate through an additional angle DBC in order for the Moon to be on the observer's meridian again. The time taken for this extra rotation is approximately 50 minutes, i.e. the lunar day is about 1d. 0h. 50m. but varies between 1d. 0h. 38m. and 1d. 01h. 06m.

The expression for the hour angle of a heavenly body applies equally well to the Moon ana planets as to a star.

G. H. A. Moon = G. H. A. Aries + S. H. A. Moon

Since the S. H. A. of the Moon can be predicted, it is combined with the G. H. A. Aries and is tabulated in the Nautical Almanac as the G. H. A. Moon for every hour G. M. T.

Similarly: G. H. A. planet = G. H. A. Aries + S. H. A. planet

and these two quantities are combined to give G. H. A. planet which is tabulated in the Nautical Almanac for each hour G. M. T. Only the planets bright enough to be observed are used - Venus, Jupiter, Saturn and Mars.

VII
ALTITUDES

THE SEXTANT AND SEXTANT ALTITUDES

The principle of the sextant is based upon two simple laws of optics:-

1. When a ray of light is reflected in a plane mirror, the angle of incidence is equal to the angle of reflection, both measured from the normal.

2. The angle of incidence, the angle of reflection and the normal all lie in the same plane.

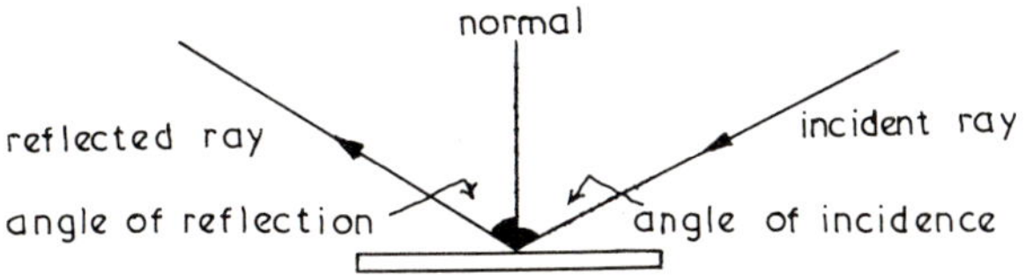

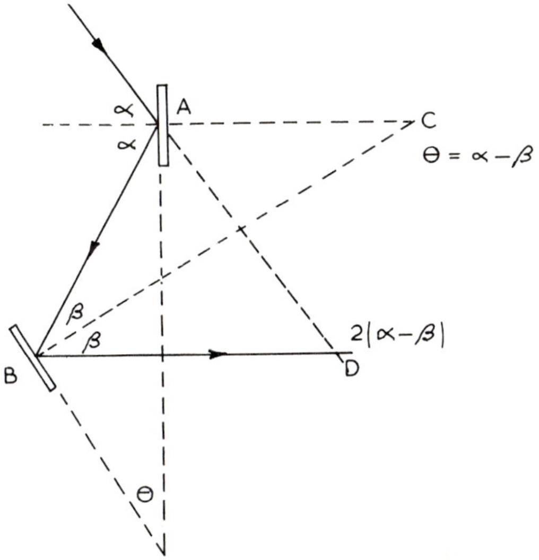

In the second figure, A and B represent two plane mirrors in the same plane. The full line indicates a ray of light doubly reflected by the two mirrors.

Let the first angle of incidence be α and the second angle of reflection be β.

The angle between the mirrors = angle between their normals = θ

In the triangle ABC;

Since the exterior angle of a triangle is equal to the sum of the interior opposite angles then angle ACB = $(\alpha - \beta)$

and in triangle ABD angle BDA = $2(\alpha - \beta)$

therefore angle BDA = 2. ACB

and the angle between the first incident ray and the second reflected rayis twice the angles between the mirrors.

The sextant is so called because its graduated arc is about one-sixth of a circle. Since the angle between the first and last directions of the ray of light is double the angle between the mirrors, the arc is calibrated to 120°.

The sextant is used for measuring the angle subtended at the observer by two distant objects. When sailing coastwise, it is used for measuring the horizontal angle between two conspicuous shore objects or the vertical angle of lighthouses etc. For celestial navigation, the sextant is used for measuring the altitude of a heavenly body - the angle, in this case, is the angle subtended at the observer, by the body and the visible horizon, measured in the plane of the vertical circle through the body and the observer's zenith.

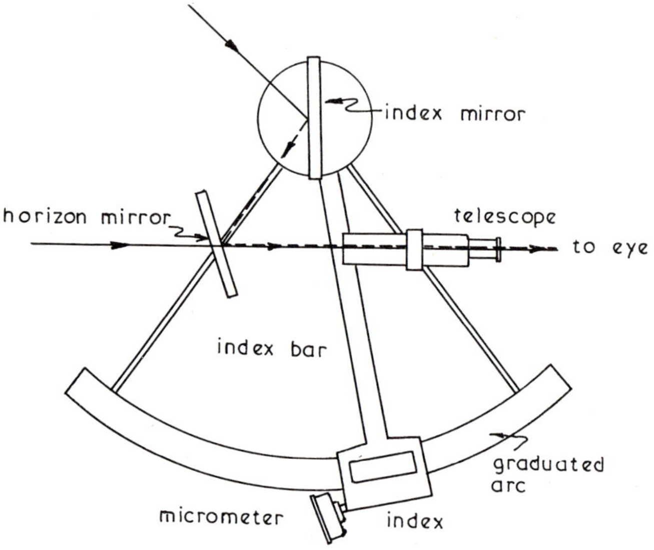

OUTLINE DIAGRAM OF THE SEXTANT

The diagram shows a sketch of the marine sextant, of which there are two types: the micrometer sextant and the practically obsolete vernier sextant. The only difference between the two types, is that the micrometer can be read with speed and accuracy without artificial aids and in poor light.

There are four adjustable errors which can be detected and eliminated by the observer, and the correct order of making the adjustments is:

1. Perpendicularity - the index mirror must be perpendicular to the plane of the instrument.

   Method: Set the index near the centre of the arc and view the true and reflected images of the arc. If they are not coincidental, turn the small screw at the side of the index mirror.

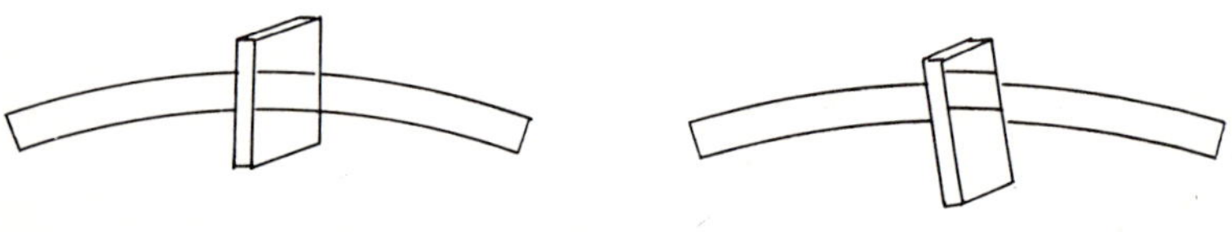

PERPENDICULARITY

2. Side error - the horizon mirror must be set perpendicular to the plane of the instrument.

Method: Fit the inverting telescope and with the index at zero observe the true and reflected images of a star, the plane of the instrument being vertical. Move the index so that the reflected image passes over the true image. If it does not pass exactly over, side error exists and can be eliminated by adjusting the screw at the side of the horizon mirror.

Alternative method: Set the index at zero and view the true and reflected images of the horizon. Tilt the sextant slightly and if the images separate then side error exists. Its correction has already been explained.

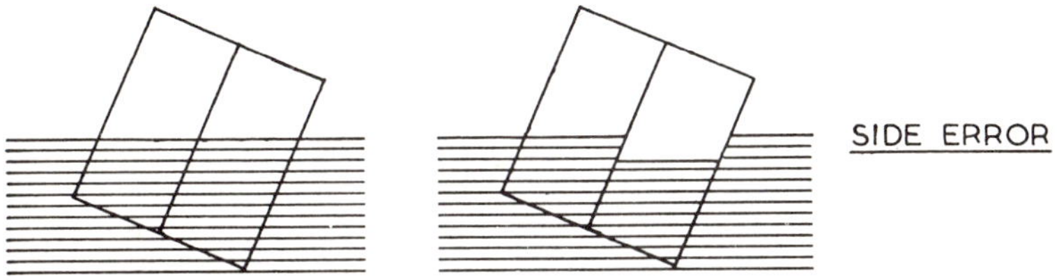

3. Index error - the horizon and index mirrors must be parallel when the index is set at zero.

Three methods for finding the index error are now given, in order of accuracy. It is pointed out, that where the index error is small (3' or less) it is better to apply it arithmetically to all readings. Continuous use of the adjusting screws may loosen them, causing inconsistent and inaccurate readings.

Method 1: Set the index a few minutes from zero, then observe a star and bring the true and reflected images into exact coincidence. The reading is then the index error. If it is necessary to correct it, then adjustment of the screw at the top or bottom of the horizon mirror will eliminate the error.

Caution - with micrometer sextants, there is always the chance of an error from "back-lash". When determining the index error by means of a star, the micrometer drum should be rotated first in one direction and then in the opposite direction. If there is any difference in the two readings, it is due to "backlash". In order to eliminate this error, all readings should be taken by rotating the drum in opposite directions and adopting the mean of the two readings.

Method 2: Fit the shade head to the inverting telescope and bring the direct and reflected limbs of the Sun into exact contact and note the reading. Then bring the reflected image of the Sun across the direct image until the opposite limbs are in exact contact and again note the reading. Half the difference between the readings gives the index error which is negative if the greater reading is 'on the arc' and positive if the greater reading is 'off the arc'.

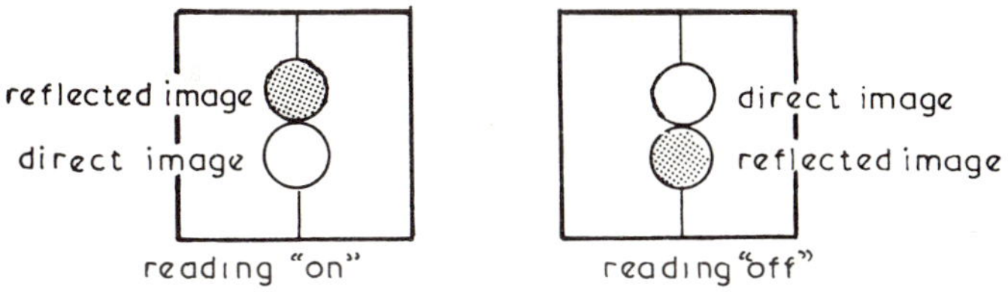

A check on the accuracy of the reading is obtained if the sum of the readings is divided by four. This gives the semi-diameter of the Sun which may be checked against the tabulated value for that day in the Abridged Nautical Almanac.

Method 3: The true and reflected images of the horizon are brought into coincidence. The reading is then the index error, if any.

4. Collimation error - the axis of the telescope, when in position, should be parallel to the plane of the instrument.

   Method: The inverting telescope is shipped with the crosswires parallel to the plane of the instrument. Two stars, not less than 90° apart, are brought into accurate contact on one wire. The sextant is then moved slightly until the bodies are on the other wire. If the two stars separate, then collimation error exists. The error can be adjusted by means of the two screws on the collar of the telescope.

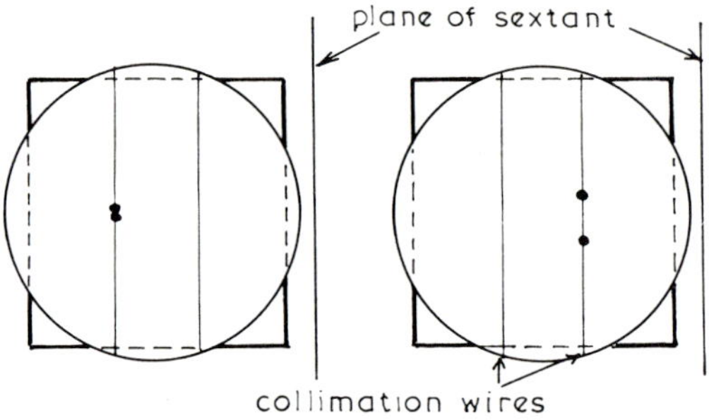

In most modern instruments, the collimation adjustment screws are no longer incorporated in the instrument, so that, if collimation error exists, it suggests that the frame of the sextant, or the collar of the telescope has been bent.

There are also a number of errors that cannot be corrected by the observer:

Centring error: this exists when the pivot of the index bar is not at the centre of the arc. The error from this cause changes in amount over various parts of the arc.

Shade errors: these occur when the sides of the individual shades are not ground parallel and results in lateral displacement of the image viewed through the defective shade. If this is suspected, it is advisable to use the shade head on the telescope which cannot produce any error since both true and reflected images pass through the same shade.

Prismatic errors in both mirrors and lenses of the telescopes.

Worm and rack errors. Peculiar to micrometer sextants, the errors vary in amount and sign. A new sextant which has not been 'run in' is liable to small and progressive changes which may make it advisable to have it re-calibrated after a period of time.

SEXTANT PARALLAX

When the sextant is used for measuring the angle subtended by two separate objects the angle actually measured is the true angle plus a small quantity known as sextant parallax.

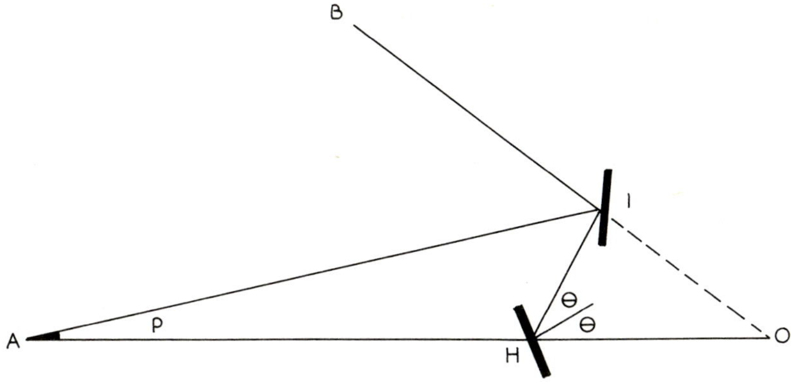

SEXTANT PARALLAX

The angle measured is BOA

The angle required is BIA

angle BIA being an external angle of the triangle AIO, then

$$\begin{aligned} BIA &= 2\ (\text{angle between mirrors}) + p \\ &= \text{angle on arc} + p \end{aligned}$$

by applying the sine rule to triangle AIH

$$\frac{IH}{\sin p} = \frac{IA}{\sin IHA} \qquad \text{and } IHA = (180^{\circ} - 2\theta)$$

and $$\sin p = \frac{IH. \sin IHA}{IA} = \frac{IH.\ \sin 2\theta}{IA}$$

this shows that the sextant parallax decreases as the distance of A increases.

In practice, sextant parallax is negligible unless A, the object viewed directly in the telescope, is close to the observer. Therefore, to reduce parallax, the telescope should always be pointed at the more distant object.

Always test for perpendicularity, side and index errors before taking sights and remove where necessary.

Use the inverting telescope for sun altitudes and small vertical angles. Use of this telescope requires a certain amount of practice, but the accuracy of the results amply repays the time spent.

Always lift the sextant by the frame or handle - NEVER by the arc or index arm.

Always keep the mirrors and telescopes clean and free from moisture.

Always read very small angles both 'on' and 'off' the arc.

Always treat the sextant with great care and avoid lending the sextant to other people - it has happened that a sextant was borrowed and looked the same afterward - but wasn't!

It is a good idea to give the worm and rack a light coating of oil, particularly if the instrument is being put away for some time.

OBTAINING THE TRUE ALTITUDE

As can be seen from the diagram, corrections must be applied to the observed altitude of a heavenly body in order to obtain its true altitude.

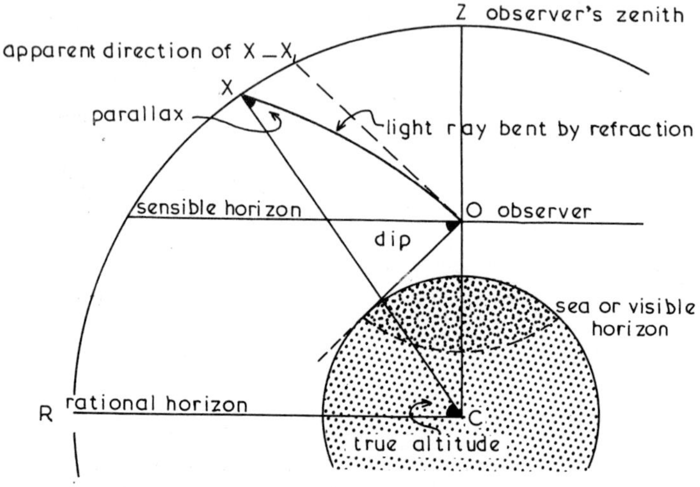

The true altitude of a celestial body is the angle it makes with the celestial, or rational, horizon, measured in a vertical plane through the body and the observer's zenith, angle XCR.

The sextant altitude: i.e. the angle actually measured between the body and the visible horizon, must have the following corrections applied to it: Index error; Dip; Refraction; Semi-diameter and parallax in that order.

Index error is discussed on page 63 and is, of course, inherent only in a particular instrument. After correction for index error, the result is known as the observed altitude.

Dip is the angle between the horizontal plane through the observer's eye (sensible horizon) and the apparent direction of the visible horizon.

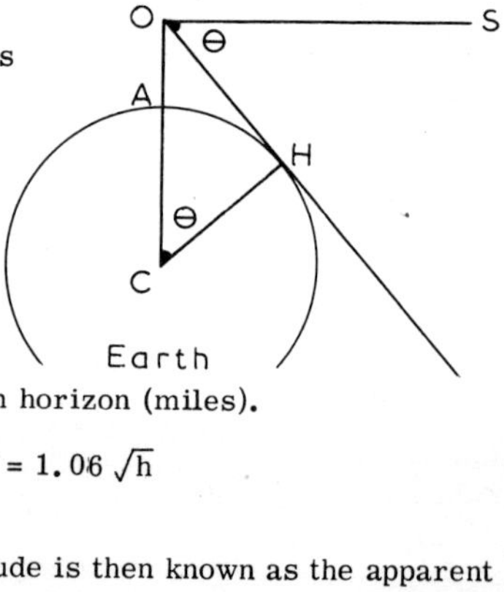

Arc AH = θ'
Angle SOH = dip in minutes

In triangle COH

Angle OHC = 90°
Angle OCH = θ'
and Angle COH = $(90^{\circ} - \theta)$
but Angle SOC = 90°

therefore Angle SOH = θ'
and dip of the sea horizon (minutes) = distance from horizon (miles).

Dip is also given by the approximate formula: $Dip' = 1.06\sqrt{h}$
where h is the height of the observer's eye in feet.

After correction for dip, the observed altitude is then known as the apparent altitude.

Refraction, or deviation of light rays from their path, occurs when rays of light pass from one medium to another of different density, or when the medium through which they are passing gradually changes density.

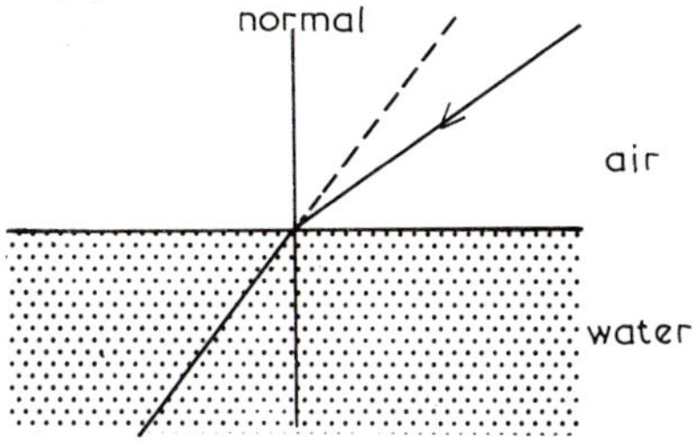

This diagram shows the effect of a ray of light passing from air to water, the refracted ray being bent towards the normal. It follows, then, that a ray passing from a dense medium to a less dense one, will be refracted away from the normal.

The diagram below shows a ray of light from some body X passing through air of gradually increasing density. The eye, which cannot see round corners, sees the body at X^1. The angular difference between XOH and X^1OH (the apparent altitude) is the angle of refraction, and is always subtracted from the apparent altitude.

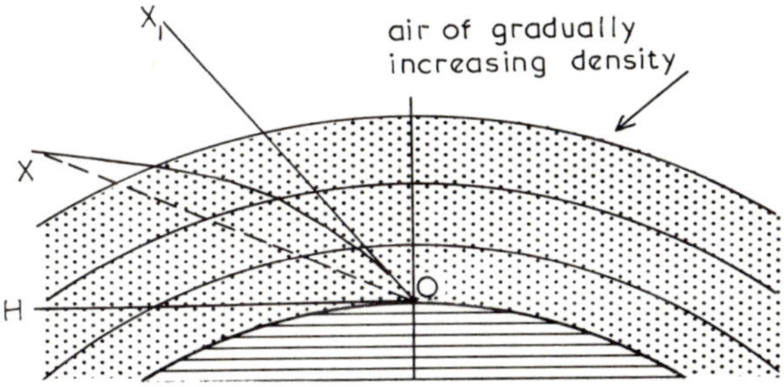

Refraction varies with the altitude of the body, being greatest when the body is on the horizon. It disappears when the body is on the observer's zenith, since the rays of light enter the Earth's atmosphere along the line of the normal.

A further effect of refraction is shown in the diagram. The horizon an observer actually sees is not a tangent from his eye to the Earth"s surface. The point that he actually sees is H^1, and since the ray is bent by refraction, the visible horizon appears in the direction OR. The dip, therefore is the angle SOR, which is less than the theorectical angle SOH. The Abridged Nautical Almanac gives values for the angle of dip which includes the effect of terrestrial refraction.

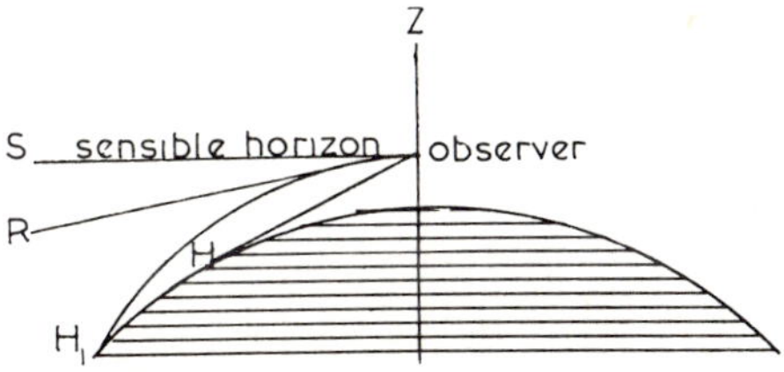

An approximate expression for Dip (including the effect of refraction) is $1.15\sqrt{h}$ Where h is the height of the observer's eye.

Marked changes can occur in refraction with abnormal temperatures and barometric pressures. Burton's Tables, for instance, quote the Mean Refraction for pressure and temperature of 29.6" and 50°F, respectively, with additional tables for when the baro. and thermometer stand at other values. Nevertheless, when abnormal refraction is suspected, all observations taken under these conditions should be treated with caution.

The correction for semi-diameter becomes necessary with altitudes of the Sun and Moon. Because of their proximity to the Earth and their relative sizes, the observer sees them as bodies having definite diameters. The altitude is measured from the upper or lower edge (limbs) and the horizon. Consequently, a correction for the semi-diameter must be applied, in order to obtain the altitude of the centre of the body.

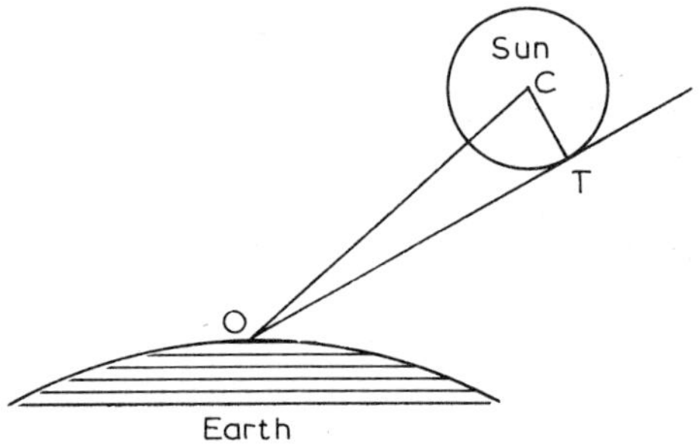

The sun's semi-diameter is measured by the angle TOC in the diagram, where OT is tangential to the surface of the Sun and therefore, at right angles to the radial line CT. CT has an approximate value of 430,000 miles and the mean distance of the centre of the Earth's centre from that of the Sun, is approximately 93,000,000 miles.

$$\text{Therefore } \sin(TOC) = \frac{CT}{OC} = \frac{430,000}{93,000,000}$$

from which, angle TOC = 16'

The Abridged Nautical Almanac gives the value of the Sun's semi-diameter for each day. It varies from 16'18" early in the year, when the Earth is closest to the Sun, to 15'45" in July, when the Earth is at its greatest distance.

The semi-diameter of the moon varies between 16.7' and 14.7' but owing to the proximity of the Earth and Moon, the semi-diameter seen by the observer actually increases as the altitude increases. This is shown in the figure. Since OM is a greater distance than OM^1, the moon's semi-diameter must be greater at M^1 than at M.

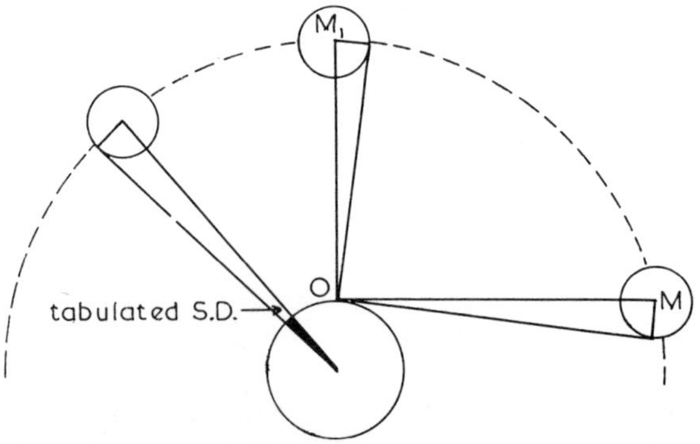

The values of the Moon's semi-diameter tabulated in the Abridged Nautical Almanac are for the angle subtended at the centre of the Earth by the Moon's radius. This value should be increased by an amount known as the augmentation of the Moon's semi-diameter. The maximum value of this augmentation is 0.31' and this occurs when the apparent altitude is 90° and when the Moon semi-diameter is maximum. This correction is applied only when extreme accuracy is required.

An additional correction is incorporated in the total correction tables for the Sun published in the Nautical Almanac, to cover the effect of irradiation between the upper limb and the horizon. The amount is 1.2' and is subtracted. Irradiation occurs with extreme contrasts of light and shade and the figure illustrates its effect.

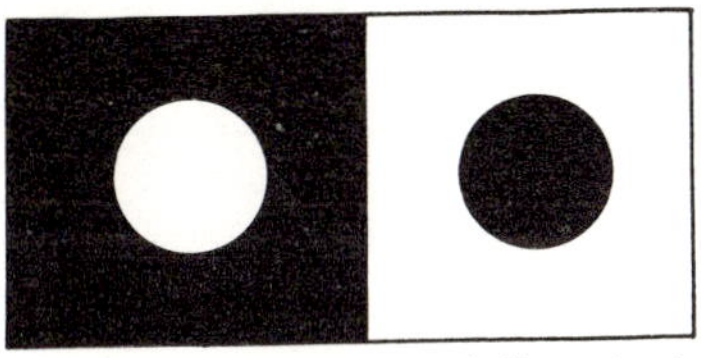

Both circles are of the same diameter, yet the circle with the dark background appears larger due to irradiation.

In the Nautical Almanac the correction for irradiation is used only for the upper limb since the Sun is then viewed against the dark background of the Earth's surface.

The correction for parallax takes into account that the true altitude must be measured at the centre of the Earth. Since stars are such vast distances from the Earth, then their altitudes at the centre of the Earth are the same as at the surface. In the case of the Sun, Moon and other members of the Solar System, parallax has definite values.

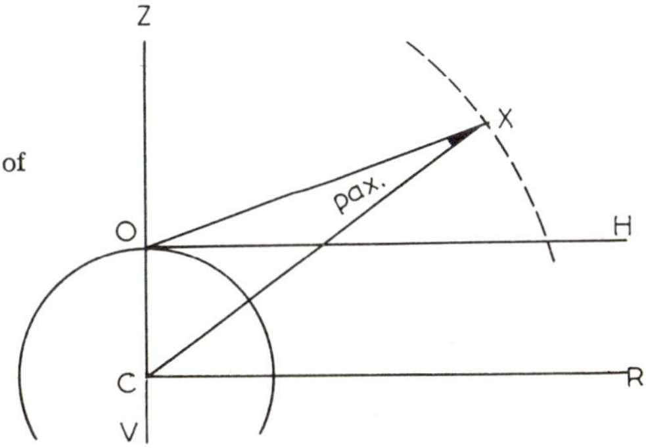

In the figure, the apparent altitude of the Moon is angle HOX and the true altitude is angle RCX.

From triangle OCX:

Angle VCX = Angle COX + Angle OXC (ext. angles of triangle)
and angle RCX + 90° = " HOX + 90° + angle OXC
therefore angle RCX = " HOX + angle OXC

i.e. true altitude = apparent altitude + parallax in altitude (OXC)

By applying the sine rule to triangle OCX:

$$\frac{\sin \text{OXC (parallax)}}{\text{OC}} = \frac{\sin \text{COX}}{\text{CX}}$$

therefore $\sin \text{(parallax)} = \frac{\text{OC}}{\text{CX}} \sin \text{COX}$

$$= \frac{\text{OC}}{\text{CX}} \sin (90^o + \text{apparent alt.})$$

$$= \frac{\text{OC}}{\text{CX}} \cos. \text{ apparent altitude.}$$

Since the cosine is zero when the altitude is 90°, it follows that there is no parallax when the body is on the observer's zenith.

Again, the cosine has maximum value when the altitude is zero, so that when the body is on the horizon, parallax will be greatest, and is then known as Horizontal Parallax (H. P.).

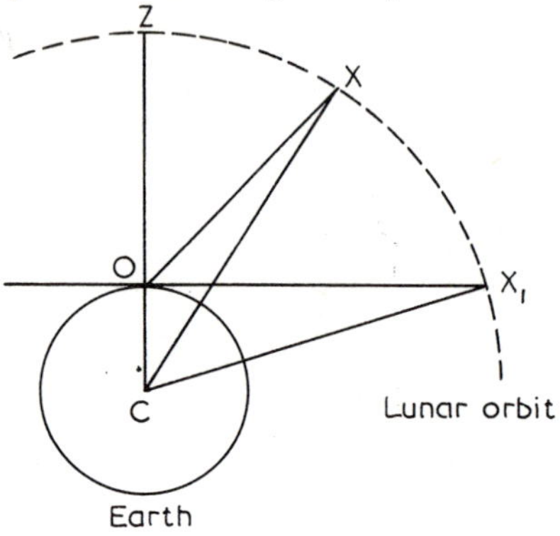

therefore sin OXC = sin (H. P.) cos (alt)

and sin (par. in alt.) = sin (H. P.) cos (alt)

since both parallax in altitude and Horizontal Parallax are very small angles, the formula becomes;

par. in alt. = H. P. cos (alt).

This formula does not take into consideration the spheroid shape of the Earth. The reduction of the Moon's H. P. is given in most navigational tables and its maximum value is never greater than 0.2'.

The tabulated values, in the Abridged Nautical Almanac, are the angles, subtended at the Moon's centre, of the equatorial radius of the Earth.

The H. P. of any body is easily calculated if the Earth's radius and the distance of the body from the Earth, are known:

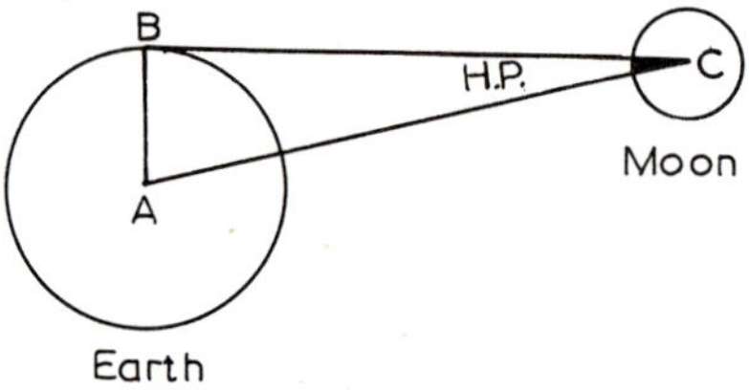

AB = Earth's radius = 4,000 miles approx.

AC = distance of Moon = 240,000 miles approx.

$$\sin ACB = \frac{4,000}{240,000} = \frac{1}{60}$$

therefore ACB = 57.3' = Moon's Horizontal Parallax.

When the Moon has some definite altitude above the horizon, its parallax must be some amount less than H. P. Its value is known as parallax in altitude.

CORRECTION OF SEXTANT ALTITUDES

In practice it is advisable to correct sextant altitudes from the tables given in the Abridged Nautical Almanac as these incorporate minor variations from the total correction tables published in the older nautical tables.

After applying the index error, if any, the observed altitude is corrected for the height of the observer's eye to give the apparent altitude. The total correction for the particular body is then applied.

The correction tables in the Abridged Nautical Almanac are based on a temperature of 10°C. (50°F) and a barometric pressure of 1010 mbs. (29.83"). In the case of the Sun an additional correction of -1.2' is included for the effect of irradiation of upper limb and horizon.

The navigational planets are treated in the same way as stars, in that the total correction is only for refraction. There are small additional corrections for Mars and Venus that take into account parallax and phase, but these corrections should be ignored for daylight observations.

In conditions other than the adopted standards of temperature and pressure additional corrections are given for the increased refraction, but these need only be used in extreme conditions and for low altitudes. Observations taken under these conditions should be treated with the greatest caution.

In each of the following examples the true altitude is required.

EXAMPLE 32

On January 30th 1958 the sextant altitude of the Sun's lower limb was 36° 54'.2 . Index error 2'.0 on the arc. Height of eye 46 feet.

	From Nautical Almanac		From Norie's Tables
Sextant altitude	36° 54'.2	. . . .	36° 54'.2
Index error	- 2'.0	. . . .	- 2'.0
Observed altitude	36° 52'.2	. . . .	36° 52'.2
Dip	- 6'.6	. . . .	6'.65
Apparent altitude	36° 45'.6	. . . .	36° 45'.55
		refraction . . .	- 1'.28
		parallax . . .	+ 0'.12
Total correction	+ 15'.0	semi-diameter . .	+ 16'.26
True altitude	37° 00'.6	. . . .	37° 00'.65

EXAMPLE 33

On 1st February 1958, the sextant altitude of the Sun's lower limb was 25° 10'.0. Index error + 1'.5, height of eye 10 feet. Barometer 30.5". Temperature 68°F.

	From Nautical Almanac		From Norie's Tables
Sextant altitude	25° 10'.0	. . . .	25° 10'.00
Index error	+ 1'.5	. . . .	+ 1'.50
Observed altitude	25° 11'.5	. . . .	25° 11'.50
Dip	- 3'.1	. . . .	- 3'.10
Apparent altitude	25° 08'.4	. . . .	25° 08'.40
		refraction . . .	- 2'.02
		temperature . . .	- 0'.09
Total correction	+ 14'.2	pressure . . .	+ 0'.06
Additional correction		parallax . . .	+ 0'.14
for tem. and press.	+ 0'.1	semi-diameter . .	+ 16'.26
True altitude	25° 22'.7	. . . .	25° 22'.75

EXAMPLE 34

1st July 1958, the sextant altitude of the Sun's upper limb was 47° 32'.0. Index error 1'.5 off the arc, height of eye 53 feet.

	From Nautical Almanac		From Norie's Tables
Sextant altitude	47° 32'.0	. . . .	47° 32'.0
Index error	+ 1'.5	. . . .	+ 1'.5
Observed altitude	47° 33'.5	. . . .	47° 33'.5
Dip	- 7'.1	. . . .	- 7'.14
Apparent altitude	47° 26'.4	. . . .	47° 26'.36
		refraction . . .	- 0'.87
		parallax . . .	+ 0'.11
Total correction	- 17'.9	semi-diameter . .	- 15'.76
True altitude	47° 08'.5	. . . .	47° 09'.83

The large difference in this case would be due to the irradiation correction in the N.A.

EXAMPLE 35

The observed altitude of Sirius was 46° 27'.4. Height of eye 30 feet.

	From Nautical Almanac		From Norie's Tables	
Sextant altitude	46° 27'.4	. . . .	46° 27'.4	46° 27'.4
Dip	- 5'.3	. . . .	- 5'.37	
Apparent altitude	46° 22'.1	. . . .	46° 22'.03	
Correction	- 0'.9	. . . .	- 0'.91	tot.corr. 6'.3
True altitude	46° 21'.2		46° 21'.12	46° 21'.1

EXAMPLE 36

1st December 1958 in latitude 32°N at 1200 G.M.T. the sextant altitude of the Moon's upper limb was 32° 15'.0. Index error - 2'.5. Height of eye 30 feet. (H.P. 56'.7 S.D. 15'.5)

	From Nautical Almanac		From Norie's Tables
Sextant altitude	32° 15'.0		32° 15'.00
Index error	- 2'.5		- 2'.50
Observed altitude	32° 12'.5		32° 12'.50
Dip	- 5'.3		- 5'.37
Apparent altitude	32° 07'.2		32° 07'.13
1st correction	+ 57'.9	ref.	- 1'.51
2nd correction	+ 3'.0	pax.	+ 48'.03
corr. for limb	- 30'.0	s'dia	- 15'.64
True altitude	32° 38'.1		32° 37'.61

semi-diameter	15'.5
augmentation	+ '.14
semi-diameter	15'.64

pax. = H.P. x cos.app.alt
= 56.7 x cos 32°.1
= 48'.03

obs.alt.	32° 12'.5
total corr. +	21.1
H.E. +	4.4
true alt.	32° 38.0

VIII

RISINGS, SETTINGS and AZIMUTHS

SUNRISE AND SUNSET

Theoretical rising or setting occurs when the centre of the heavenly body is on the observer's rational or celestial horizon, to the eastward or westward of his meridian. The true zenith distance is then 90°.

Visible rising or setting occurs when the upper limb of the body is just on the observer's visible horizon. The observed altitude of the body is then 00°00'.

Assuming no height of eye and Sun's semi-diameter for the day is 16'

Observed altitude	00°00.0'
Refraction	- 34.0'
Semidiameter	- 16.0'
True altitude	- 00°50.0'
Zenith distance	90°50.0'

It can be seen that at visible sunrise or sunset the Sun's centre is about 1° below the celestial horizon, and therefore visible sunrise occurs before theoretical sunrise, and visible sunset after theoretical sunset.

Considering the Moon -

Observed altitude	00°00.0'
Refraction	- 34.0'
Semidiameter	- 16.0'
Average parallax	+ 57.0'
True altitude	00°07.0'
Zenith distance	89°53.0'

Therefore, theoretical and visible moonrise and moonset very nearly occur at the same moment.

The times of visible sunrise and sunset are tabulated on the daily pages of the Abridged Nautical Almanac. These tables solve the PZX triangle for the angle P when ZX is 90°50'. The tables give the exact L. M. T. of the phenomena on the Greenwich meridian over a considerable range of latitudes. It is necessary to interpolate for latitude and a table assisting this operation is printed at the back of the Almanac. The actual times quoted are for the middle day of the page and again interpolation may be necessary for the given date. The small variation for longitude has no practical significance, and it is customary to regard the time given in the tables as the L. M. T. for any meridian.

Twilight is that period of the day when the observer is still receiving some scattered light from the Sun, although the body is below the horizon. Twilight is divided into three stages:

<u>CIVIL TWILIGHT</u>: said to begin or end when the centre of the Sun is 6° below the horizon. True Zenith Distance is 96°.

<u>NAUTICAL TWILIGHT</u>: said to begin or end when the centre of the Sun is 12° below the horizon. True Zenith Distance is 102°.

<u>ASTRONOMICAL TWILIGHT</u>: said to begin or end when the centre of the Sun is 18° below the horizon, at which moment complete darkness is assumed to begin or end as far as the Sun is concerned. True Zenith Distance is 108°.

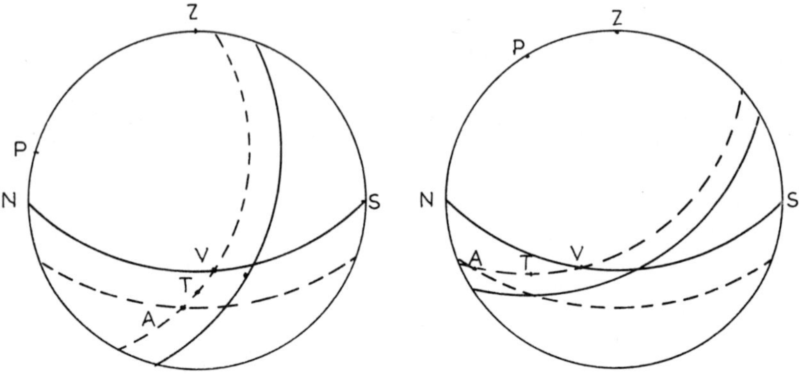

The figures show the relative positions of the Sun at visible (v), theoretical sunrise (T) and at the end of twilight (A). To an observer in low latitudes the apparent path of the Sun very nearly approaches a perpendicular with the horizon therefore twilight is of relatively short duration. If the observer is in high latitudes, the angle between the Sun's path and the horizon is more acute and twilight lasts for a longer period of time.

If the circle of declination does not fall 18° below the horizon twilight is continuous throughout the night. This occurs when the latitude and declination have the same name and their sum is greater than 72°. The limits for civil and nautical twilight are obtained by substituting 6° and 12° for 18°.

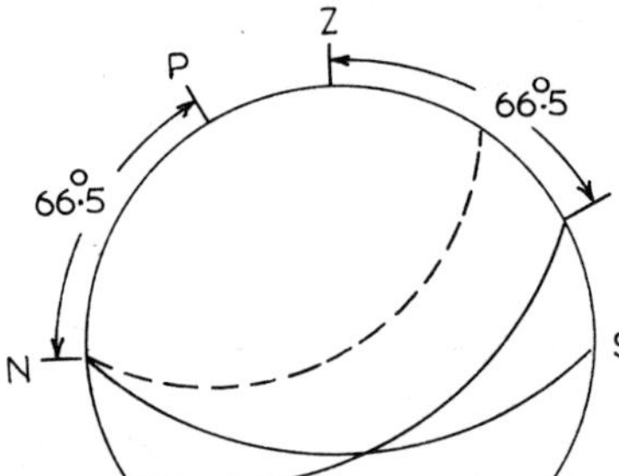

If the circle of declination just touches the horizon, then the Sun cannot set. In the northern hemisphere the limiting latitude for this phenomenon to occur is (90° - max. northerly declination) i.e. $90° - 23\frac{1}{2}° = 66\frac{1}{2}°$. In this latitude the Sun will remain above the horizon all night on one occasion only during the year.

At the same time astronomical twilight will last all night on the same night in $(90° - 23\frac{1}{2}° - 18°)$, i.e. latitude $48\frac{1}{2}°$N. These limits also apply to the southern hemisphere, with their names changed to south.

CIRCUMPOLAR BODIES

Although every heavenly body, to an observer on the Earth, describes a circle about the celestial pole, the term circumpolar, when applied to a heavenly body, denotes that it does not set and is always above the observer's horizon.

MOONRISE AND MOONSET

To avoid a long and laborious calculation of the times of moonrise and moonset, the times are tabulated in the Abridged Nautical Almanac for an observer on the meridian of Greenwich with no height of eye.

The L. M. T. of moonrise and moonset at Greenwich must be corrected for latitude and then a correction for the longitude of the observer must be applied.

This correction is simply a proportion of the daily difference of moonrise or moonset for the given latitude:

$$\text{correction for longitude} = \frac{\text{longitude in degrees}}{360} \text{ x daily difference}$$

and the rule for applying this correction:

Observer in west longitude:- take daily difference between day in question and following day: correction to be added.

Observer in east longitude:- take daily difference between day in question and the preceding day: correction to be subtracted.

Tables to facilitate the interpolation for latitude and to obtain the longitude correction are printed at the end of the Abridged Nautical Almanac. The reader is also reminded that the same procedure for longitude correction must be carried out when working the L. M. T. of the Moon's meridian passage. The exception to this rule may sometimes occur when the observer is in high latitudes and the times of moonrise or moonset may get earlier each day. The sign of the longitude correction must then be changed.

EXAMPLE 37

Find the L.M.T. of moonrise and moonset on the 19th October 1958 for two observers. One in latitude 42°00'N., longitude 135°00'E. the other in latitude 72°00'N., longitude 135°00'E.

	Moonrise		Moonset	
L.M.T. Greenwich	18d.12h.04m.	19d.12h.50m.	18d.22h.23m.	19d.23h.23m.
lat. correction	+ 05m.	+ 04m.	– 05m.	– 04m.
L.M.T. lat 42°N.	18d.12h.09m.	19d.12h.54m.	18d.22h.18m.	19d.23h.19m.
daily difference	45m.		61m.	
long.correction	– 17m.		– 23m.	
L.M.T. lat.42°N.	19d.12h.54m.		19d.23h.19m.	
long.correction	– 17m.		– 23m.	
L.M.T. observer	19d.12h.37m. moonrise		19d.22h.56m. moonset	

Observer in lat.72°N. long.135°E.

	Moonrise		Moonset	
L.M.T. Greenwich	18d.16h.24m.	19d.16h.01m.	18d.18h.08m.	19d.20h.19m.
lat.correction	-	-	-	-
L.M.T. lat 72°N.	18d.16h.24m.	19d.16h.01m.	18d.18h.08m.	19d.20h.19m.
daily difference	23m.		131m.	
long.correction	+ 9m.		– 49m.	
L.M.T. lat. 72°N.	19d.16h.01m.		19d.20h.19m.	
long.correction	+ 09m.		– 49m.	
L.M.T. observer	19d.16h.10m. moonrise		19d.19h.30m. moonset.	

AMPLITUDES

In the figure, drawn on the plane of the rational horizon, the triangle PZX is the astronomical triangle for a celestial body that is on the rational horizon to the east of the observer, i.e. it is rising.

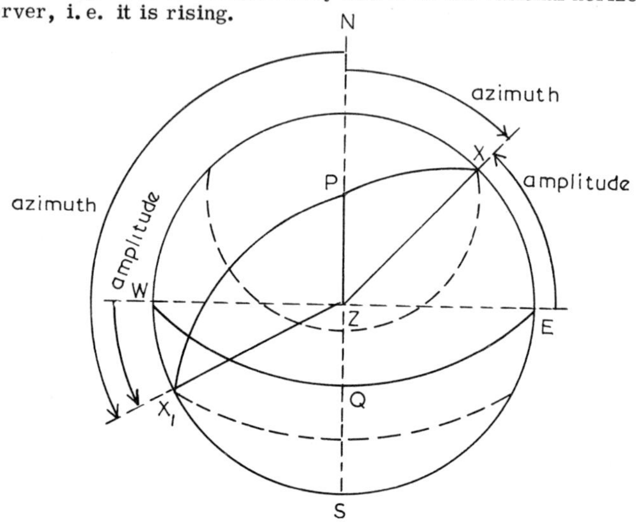

The angle PZX is the azimuth and angle EZX is the rising amplitude.

The triangle PZX_1 is the astronomical triangle for a celestial body that is on the rational horizon to the west of the observer, i.e. it is setting.

The angle PZX_1 is the azimuth and the angle WZX_1 is the setting amplitude.

The amplitude of a rising body is named East followed by the name of the declination. E.g. E.10°N.

The amplitude of a setting body is named West followed by the name of the declination. E.g. W.15°S.

Since the triangle is quadrantal (zenith distance ZX is 90°) the amplitude may be found by using Napier's Rules:-

$$\begin{aligned}
\sin \text{co. PX} &= \cos \text{co. PZ} . \cos Z \\
\cos PX &= \sin PZ . \cos Z \\
\cos (90^\circ - \text{dec}) &= \sin (90^\circ - \text{lat}) . \cos (90^\circ - \text{amplitude}) \\
\sin \text{declination} &= \cos \text{latitude} . \sin \text{amplitude}
\end{aligned}$$

therefore: <u>sin amplitude = sin declination. sec latitude.</u>

EXAMPLE **38**

The sun rose bearing 124° by compass. Given that the latitude was 39°50'N. and the declination was 19°47'S. find the compass error.

lat 39°50' log sec	= 0.11469				
dec 19°47' log sin	= 9.52951				
log.sin true amp.	= 9.64420	true amplitude	= E.26°09'.1S.		
		observed amplitude	= E.34°00'.0S.		
		compass error	= 7°50'.9W.		

Most navigational tables include a table of pre-computed amplitudes which afford a considerable saving in time and also offer an invaluable check for examination problems.

EXAMPLE 39

Find the true amplitude of the Sun when setting, Latitude 37°30'S. Declination 22° 15'S.

latitude	37°30'	log sec	0.10053
declination	22°15'	log sin	9.57824
log sin true amplitude			9.67877
true amplitude (calculated)			W.28°30'.5S
true amplitude (Norie's)			W.28°30'.5S

It must be remembered that the formula, and for that matter, ready use tables, depend on the true altitude of the body being 0°. In the case of the Sun, the observation should be made when the Sun's lower limb is about a semi-diameter above the horizon. This reduces the error due to parallax and refraction (see page 73). In high latitudes when the apparent path of the Sun makes an acute angle with the horizon, it will be difficult to estimate the correct moment and it is therefore better to time the observation.

AZIMUTHS

In the celestial triangle PZX, the azimuth of the body X is given by the angle Z.

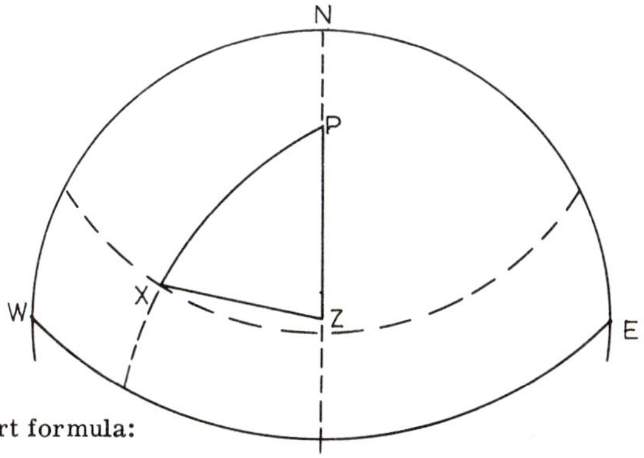

From the four part formula:

$$\cos P.\cos PZ = \sin PZ.\cot PX - \sin P.\cot Z$$

dividing throughout by sin P. sin PZ

$$\frac{\cos P.\cos PZ}{\sin P.\sin PZ} = \frac{\sin PZ.\cot PX}{\sin PZ.\sin P} - \frac{\sin P.\cot Z}{\sin P.\sin PZ}$$

$$\cot P.\cot PZ = \cot PX.\operatorname{cosec} P - \cot Z.\operatorname{cosec} PZ$$

or cot H. A. tan lat. = tan dec. cosec H. A. — cot azimuth. sec lat.

This expression is tabulated in nautical tables as the A, B, C tables.

cot H. A. tan lat. is tabulated for all values of H. A. and lat. as A,
cosec H. A. tan dec. is tabulated for similar values of H. A. and dec. as B,
cot az. sec lat. is tabulated for combinations of azimuth and lat. as C.

Procedure

Table A is entered with latitude and L. H. A. and the quantity is named opposite to the latitude in Norie's but + ve in Burton's except when the L. H. A. is greater than 6 hours (90°) when the procedure is reversed.

Table B is entered with declination and L. H. A. and the quantity is always named the same as the declination in Norie's but - ve in Burton's when latitude and declination have the same name; when they are of different name the procedure is reversed.

To find the quantity C :- A and B same names (or signs) - add
A and B different names (or signs) - subtract
C takes name (or sign) of the larger quantity.

The following is an example of the A, B, C tables with two examples on their use.

A TABLE A — HOUR ANGLE

Lat. °	45° 315°	46° 314°	47° 313°	48° 312°	49° 311°	50° 310°	51° 309°	52° 308°	53° 307°	54° 306°	55° 305°
0	·00	·00	·00	·00	·00	·00	·00	·00	·00	·00	·00
1	·02	·02	·02	·02	·02	·01	·01	·01	·01	·01	·01
2	·03	·03	·03	·03	·03	·03	·03	·03	·03	·03	·02
3	·05	·05	·05	·05	·05	·04	·04	·04	·04	·04	·04
4	·07	·07	·07	·06	·06	·06	·06	·05	·05	·05	·05
5	·09	·08	·08	·08	·08	·07	·07	·07	·07	·06	·06
6	·11	·10	·10	·09	·09	·09	·09	·08	·08	·08	·07
7	·12	·12	·11	·11	·11	·10	·10	·10	·09	·09	·09
8	·14	·14	·13	·13	·12	·12	·11	·11	·11	·10	·10
9	·16	·15	·15	·14	·14	·13	·13	·12	·12	·12	·11
10	·18	·17	·16	·16	·15	·15	·14	·14	·13	·13	·12
11	·19	·19	·18	·18	·17	·16	·16	·15	·15	·14	·14
12	·21	·21	·20	·19	·18	·18	·17	·17	·16	·15	·15
13	·23	·22	·22	·21	·20	·19	·19	·18	·17	·17	·16
14	·25	·24	·23	·22	·22	·21	·20	·19	·19	·18	·17
15	·27	·26	·25	·24	·23	·22	·22	·21	·20	·19	·19
16	·29	·28	·27	·26	·25	·24	·23	·22	·22	·21	·20
17	·31	·30	·29	·28	·27	·26	·25	·24	·23	·22	·21
18	·32	·31	·30	·29	·28	·27	·26	·25	·24	·24	·23
19	·34	·33	·32	·31	·30	·29	·28	·27	·26	·25	·24
20	·36	·35	·34	·33	·32	·31	·29	·28	·27	·26	·25
21	·38	·37	·36	·35	·33	·32	·31	·30	·29	·28	·27
22	·40	·39	·38	·36	·35	·34	·33	·32	·30	·29	·28
23	·42	·41	·40	·38	·37	·36	·34	·33	·32	·31	·30
24	·45	·43	·42	·40	·39	·37	·36	·35	·34	·32	·31
25	·47	·45	·44	·42	·41	·39	·38	·36	·35	·34	[illegible]
26	·49	·47	·46	·44	·42	·41	·39	·38	·37	·35	[illegible]
27	·51	·49	·48	·46	·44	·43	·41	·40	·38	·37	[illegible]
28	·53	·51	·50	·48	·46	·45	·43	·42	·40	·39	
29	·55	·54	·52	·50	·48	·47	·45	·43	·42	·40	
30	·58	·56	·54	·52	·50	·48	·47	·45	·44	·42	
31	·60	·58	·56	·54	·52	·50	·49	·47	·45	·44	
32	·62	·60	·58	·56	·54	·52	·51	·49	·47	·45	
33	·65	·63	·61	·58	·56	·55	·53	·51	·49	·47	
34	·67	·65	·63	·61	·59	·57	·55	·53	·51	·49	
35	·70	·68	·65	·63	·61	·59	·57	·55	·53	[illegible]	
36	·73	·70	·68	·65	·63	·61	·59	·57	·55	[illegible]	
37	·75	·73	·70	·68	·66	·63	·61	·59	·57		
38	·78	·75	·73	·70	·68	·66	·63	·61	·59		
39	·81	·78	·76	·73	·70	·68	·66	·63	[illegible]		
40	·84	·81	·78	·76	·73	·70	·68	·66			
41	·87	·84	·81	·78	·76	·73	·70	[illegible]			
42	·90	·87	·84	·81	·78	·76	[illegible]				
43	·93	·90	·87	·84	·81	·78					
44	·97	·93	·90	·87	·84						
45	[illegible]										
46											
47											

A—Named opposite to Latitude, **except** when Hour Angle is between 90° and 270°

B TABLE B — HOUR ANGLE.

Dec. °	45° 315°	46° 314°	47° 313°	48° 312°	49° 311°	50° 310°	51° 309°	52° 308°	53° 307°
0	·00	·00	·00	·00	·00	·00	·00	·00	·00
1	·02	·02	·02	·02	·02	·02	·02	·02	·02
2	·05	·05	·05	·05	·05	·05	·04	·04	·04
3	·07	·07	·07	·07	·07	·07	·07	·07	·07
4	·10	·10	·10	·09	09	·09	·09	·09	·09
5	·12	·12	·12	·12	·12	·11	·11	·11	·11
6	·15	·15	·14	·14	·14	·14	·14	·13	·13
7	·17	·17	·17	·17	·16	·16	·16	·16	·15
8	·20	·20	·19	·19	·19	·18	·18	·18	·18
9	·22	·22	·22	·21	·21	·21	·20	·20	·20
10	·25	·25	·24	·24	·23	·23	·23	·22	·22
11	·27	·27	·27	·26	·26	·25	·25	·25	·24
12	·30	·30	·29	·29	·28	·28	·27	·27	·27
13	·33	·32	·32	·31	·31	·30	·30	·29	·29
14	·35	·35	·34	·34	·33	·33	·32	·32	·31
15	·38	·37	·37	·36	·36	·35	·34	·34	·34
16	·41	·40	·39	·39	·38	·37	·37	·36	·36
17	·43	·43	·42	·41	·41	·40	·39	·39	·38
18	·46	·45	·44	·44	·43	·42	·42	·41	·41
19	·49	·48	·47	·46	·46	·45	·44	·44	·43
20	·51	·51	·50	·49	·48	·48	·47	·46	·46
21	·54	·53	·52	·52	·51	·50	·49	·49	·48
22	·57	·56	·55	·54	·54	·53	·52	·51	·51
23	·60	·59	·58	·57	·56	·55	·55	·54	·53
24	·63	·62	·61	·60	·59	·58	·57	·57	·56
25	·66	·65	·64	·63	·62	·61	·60	·59	·58
26	·69	·68	·67	·66	·65	·64	·63	·62	·61
27	·72	·71	·70	·69	·68	·67	·66	·65	·64
28	·75	·74	·73	·72	·70	·69	·68	·67	·67
29	·78	·77	·76	·75	·73	·72	·71	·70	·69
30	·82	·80	·79	·78	·76	·75	·74	·73	·72
31	·85	·84	·82	·81	·80	·78	·77	·76	·75
32	·88	·87	·85	·84	·83	·82	·80	·79	·78
33	·92	·90	·89	·87	·86	·85	·84	·82	·81
34	·96	·94	·92	·91	·89	·88	·87	·86	·84
35	·99	·97	·96	·94	·93	·91	·90	·89	·88
36	1·03	1·01	·99	·98	·96	·95	·93	·92	[illegible]
37	1·07	1·05	1·03	1·01	1·00	·98	·97	·96	[illegible]
38	1·11	1·09	1·07	1·05	1·04	1·02	1·00	·99	
39	1·15	1·13	1·11	1·09	1·07	1·06	1·04	1·03	
40	1·19	1·17	1·15	1·13	1·11	1·10	1·08	1·06	
41	1·23	1·21	1·19	1·17	1·15	1·13	1·12	[illegible]	
42	1·28	1·25	1·23	1·21	1·19	1·18	1·16	[illegible]	
43	1·32	1·30	1·28	1·25	1·24	1·22	[illegible]		
44	1·37	1·34	1·32	1·30	1·28	1·26			
45	[illegible]	[illegible]	[illegible]	[illegible]	[illegible]				

B—Always named the **same** as Declination.

C TABLE C C

A ± B = Lat. °	·80′	·82′	·84′	·86′	·88′	A & B CORRECTION. AZIMUTHS.	1·60′	1·64′	1·68′	1·72′	1·76′	1·80′ = A ± B	Lat. °
0	51·3	50·6	50·0	49·3	48·7		32·0	31·4	30·8	30·2	29·6	29·1	0
5	51·5	50·8	50·1	49·4	48·8		32·1	31·5	30·9	30·3	29·7	29·2	5
10	51·8	51·1	50·4	49·7	49·1		32·4	31·8	31·2	30·6	30·0	29·4	10
14	52·2	51·5	50·8	50·2	49·5		[illegible]	32·1	31·5	30·9	30·4	29·8	14
18	52·7	52·1	51·4	50·7	50·1		[illegible]	32·7	32·0	31·4	30·9	30·3	18
20	53·1	52·4	51·7	51·1	[illegible]		[illegible]	33·0	32·4	31·8	31·2	30·6	20
22	53·4	52·8	52·1	51·4	50·8			33·3	32·7	32·1	31·5	30·9	22
24	53·8	53·2	52·5	51·8	51·2			33·7	33·1	32·5	31·9	31·3	24
26	54·3	53·6	52·9	52·3	[illegible]			[illegible]	33·5	32·9	32·3	31·7	26
28	54·8	54·1	53·4	52·8	[illegible]			[illegible]	34·0	33·4	32·8	32·2	28

A & B take Same names, Sum, (add.) } RULE TO FIND C CORRECTION { A & B take Different names, Difference, (sub.)

C CORRECTION, (A ± B) is named the same as the greater of these quantities.

C **AZIMUTH** takes combined names of C Correction and Hour Angle. C

EXAMPLE **40**

In latitude 22°20'N. find the true azimuth of a star whose declination is 42°48'S. and L.H.A. 314°.

Note:- 314° = 46°E.

From Table A.			From Table B.		
lat 22°	L.H.A. 46°	A = 0.39	dec. 42°	L.H.A. 46°	B = 1.25
lat 23°	L.H.A. 46°	A = 0.41	dec. 43°	L.H.A. 46°	B = 1.30
lat 22°20'	L.H.A. 46°	A = 0.397.S	dec. 43°48'	L.H.A. 46°	B = 1.29.S

A = 0.397 S
B = 1.29 S
C = 1.687 S

From Table C.

lat 22°	C 1.68	Azimuth 32°.7	C 1.72	Azimuth 32°.1
lat 24°	C 1.68	Azimuth 33°.1	C 1.72	Azimuth 32°.5
lat 22°20'	C 1.68	Azimuth 32°.76	C 1.72	Azimuth 32°.16
lat 22°20'	C 1.687	Azimuth 32°.65		

True azimuth = S32°.65E or 147°.35

EXAMPLE **41**

In latitude 22°15'N. the declination of a certain star was 42°15'N. and the L.H.A. was 47°30'W. Find the true azimuth.

It is essential to interpolate correctly to maintain accuracy.

From Table A.

	lat. 22°	L.H.A. 47°	A = 0.38	L.H.A. 48°	A = 0.36
	lat. 23°	L.H.A. 47°	A = 0.40	L.H.A. 48°	A = 0.38
∴	lat. 22°15'	L.H.A. 47°	A = 0.385	L.H.A. 48°	A = 0.365
and	lat. 22°15'	L.H.A. 47°30'	A = 0.375S		

From Table B.

	dec. 42°	L.H.A. 47°	B = 1.23	L.H.A. 48°	B = 1.21
	dec. 43°	L.H.A. 47°	B = 1.28	L.H.A. 48°	B = 1.25
∴	dec. 42°15'	L.H.A. 47°	B = 1.243	L.H.A. 48°	B = 1.22
and	dec. 42°15'	L.H.A. 47°30'	B = 1.232N		

A = 0.375S
B = 1.232N
C = 0.857N

From Table C

	lat. 22°	C = 0.84	Az. 52°.1	C = 0.86	Az. 51°.4
	lat. 24°	C = 0.84	Az. 52°.5	C = 0.86	Az. 51°.8
∴	lat. 22°15'	C = 0.84	Az. 52°.2	C = 0.86	Az. 51°.5
and	lat. 22°15'	C = 0.857	Az. 51°.6		

True azimuth N51°.6W or 308°.4

For a further use of the A.B.C. tables see pages 33 and 34.

IX
POSITION LINES

POSITION LINES

The geographical position of any heavenly body can be readily determined if the observer knows three things:-

1. the declination of the body,
2. the L. H. A. of the body,
3. his own west longitude,

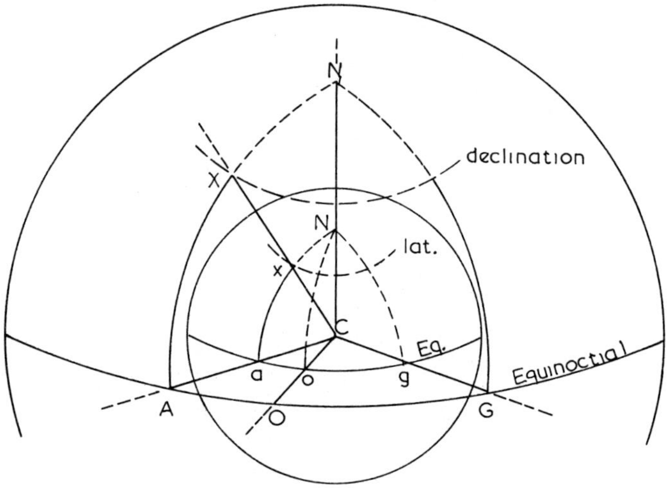

The figure represents the celestial sphere with the Earth at the centre. CX is a line from the centre of the Earth to the heavenly body, hence, where this line cuts the Earth's surface is the geographical position of the body. NCo is the plane of the observer's meridian. NCg is the plane of the Greenwich meridian.

Arc AX is the declination of the body
Arc ax is the latitude of the body
Arc OA is the L. H. A. of the body
Arc GO is the west longitude of the observer.

therefore arc GA is the G. H. A. of the body and

arc ga is the west longitude of the body.

Summarising: Latitude of a heavenly body = declination of body
West longitude of body = G. H. A. of body.

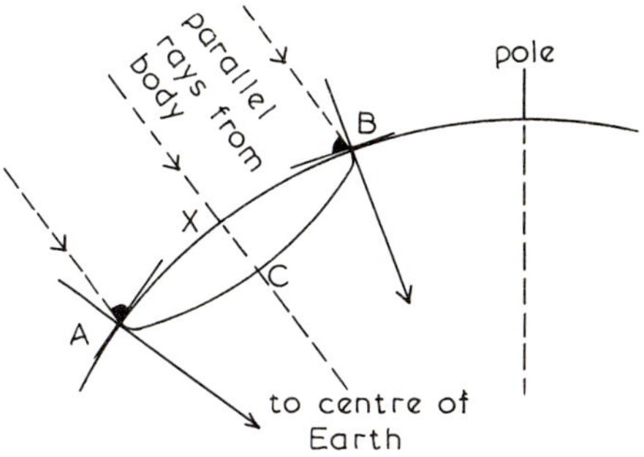

In the figure, at a certain instant a heavenly body is above geographical position X. The true altitude of this body would be the same to all observers of the body situated around the circle ABC.

This circle of 'equal altitude' has a radius in miles equal to the zenith distance of the body in minutes of arc.

Therefore an observer, having found the true altitude of a body must be situated somewhere on a circle having a radius equal to the true zenith distance of the body. If the radius of this circle is small - such as may occur in low latitudes and high altitudes - this circle could be drawn on a Mercator chart without appreciable loss of accuracy.

In general, the zenith distance will be too large to plot on the chart and so a very small arc is taken, which may be regarded as a tangent to the circle of equal altitude and consequently a straight line. This line is known as the 'position line' and is perpendicular to the bearing of the body's geographical position from the observer. If more than one position line is obtained, the intersection of them fixes the observer's position. The problem facing the navigator is then two-fold:-

1. Calculating a position that lies somewhere on the circle of equal altitude.

2. Obtaining the direction of azimuth, of the body's geographical position.

MERIDIAN ALTITUDES

The simplest method of finding the position line is to take an altitude of a body when it is on the observer's meridian. At its upper transit of the meridian it will have attained its maximum altitude to a stationary observer, and at lower transit (below the pole) the altitude will be its minimum. The azimuth is always due north or south.

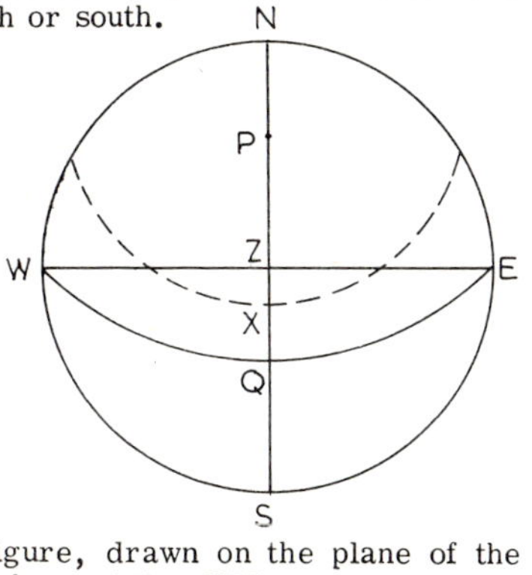

Note: the bearing of the body is to the South of the observer therefore his zenith is north of the Body.

In the figure, drawn on the plane of the rational horizon, the body is at X on the observer's meridian NZS.

The latitude of the observer is the arc QZ and is equal to the sum of the arcs ZX (zenith distance) and QX (declination).

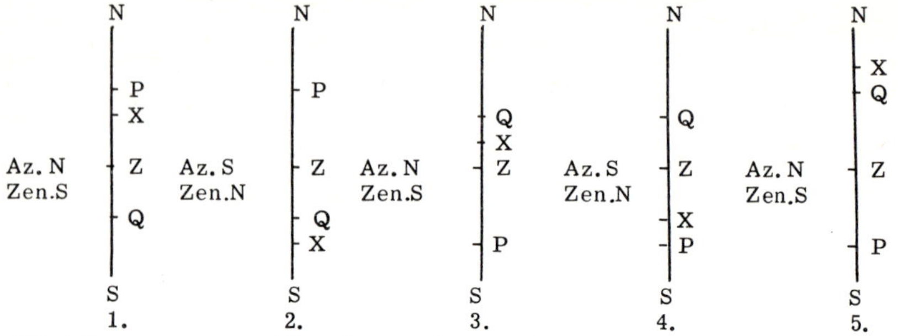

The above figures represent the observer's meridian with the following details:

1. Observer in north lat., declination same name but greater than lat.
2. Observer in north lat., declination different name, i.e. south.
3. Observer in south lat., declination same name but less than lat.
4. Observer in south lat., declination same name but greater than lat.
5. Observer in south lat., declination different name, i.e. north.

In 1. latitude (QZ) = declination (QX)- zenith distance (ZX)

2. latitude = zenith distance - declination

3. latitude = declination + zenith distance

4. latitude = declination - zenith distance

5. latitude = zenith distance - declination

the latitude taking the name of the greater quantity.

In the figure below, the body at X_1 is at its lower, or below pole transit. The problem of finding the latitude (QZ) is readily solved:-

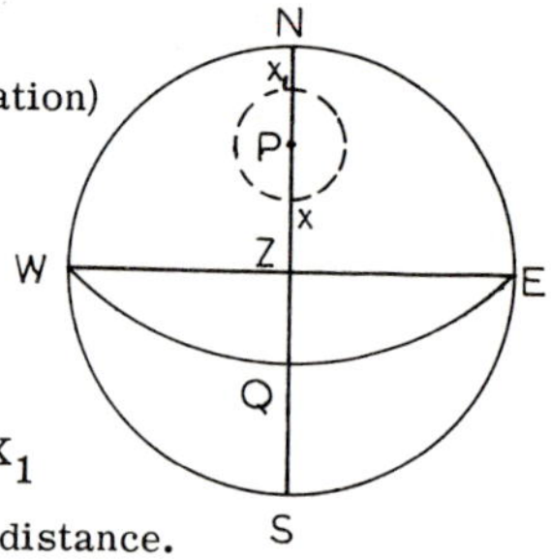

$PX = PX_1$ = polar distance of body (90° - declination)

ZX_1 = zenith distance

NX_1 = true altitude of body on lower transit

since $PQ = NZ = 90^o$ (by definition)

then $PN = QZ$ = latitude and $PN = NX_1 + PX_1$

i.e. latitude = true altitude at lower transit + polar distance.

Summarising:-

BODY AT UPPER TRANSIT

latitude = zenith distance $\pm$ declination

BODY AT LOWER TRANSIT

latitude = true altitude + polar distance

By this method, the observer obtains the latitude of the position line passing through the D. R. meridian. At the same time it must be appreciated that the position line, being perpendicular to the azimuth lies in an east/west line.

Before the meridian altitude can be calculated, it is necessary to know the declination of the body. In the case of a star this can be read directly from the Nautical Almanac. In the case of the Sun, Moon and navigational planets, it is necessary to know the G. M. T. of meridian passage in order to obtain the correct declination.

MERIDIAN PASSAGE OF THE SUN

Although it is common practice at sea to observe the meridian altitude of any celestial body when it has attained maximum altitude, this is only valid for a stationary observer. The assumption that maximum and meridian altitude occur simultaneously for a moving observer is false, and particularly if the observer's course has a large northerly component, the error can be several miles.

The correct method is to time the observation. This can be done by an approximation to the nearest minute of the time that the body will cross the meridian.

The L. M. T. of meridian passage of the Sun is given for each day in the Abridged Nautical Almanac and by applying the longitude in time, the G. M. T. of transit at the observer's position is then known.

EXAMPLE 42

Find the G.M.T. of the Sun's meridian passage to an observer in longitude 125° 35'W. (From Abridged Nautical Almanac - Mer.pass 11h.57m.)

L.M.T. Meridian passage	11h.57m.
Longitude in time	8h.22.3m.
G.M.T. passage at ship	20h.19.3m.

MERIDIAN PASSAGE OF THE MOON

As previously explained, the lunar day is nearly one hour longer than the mean solar day. A proportion of this daily difference must be taken into account when calculating the time of meridian passage of the Moon.

The L. M. T. of the Moon's upper passage at the meridian of Greenwich is given each day in the daily pages of the Abridged Nautical Almanac.

The proportion of the daily difference is given by:

$$(\text{daily difference}) \times \frac{\text{observer's longitude}}{360^\circ}$$

If the observer is in east longitude, meridian passage occurs before passage at Greenwich, therefore the daily difference between the day in question and the preceding day is taken; the proportion of this is to be subtracted from the L. M. T. at Greenwich to give the L. M. T. at ship. Correct application of the observer's longitude gives the G. M. T. at ship.

If the observer is in west longitude, meridian passage occurs after passage at Greenwich, therefore the daily difference between the day in question and the following day is taken; the proportion of this is to be added to the L. M. T. at Greenwich to give the L. M. T. at ship.

As sometimes happens there is no meridian passage at Greenwich for the given date, the daily difference is taken between the day before and the day after the date in question.

EXAMPLE 43

Find the zone time of the Moon's meridian passage for an observer in longitude 108°E. on the 21st August 1958.

From the Nautical Almanac.

Note: The times of upper passage only should be used.

Moon				
Day	Mer.pass Upper	Mer.pass Lower	Age	Phase
20	1648	0420	05d	◐
21	1744	0516	06d	
22	1839	0611	07d	

L.M.T. meridian passage at Greenwich 20d.16h.48m.
L.M.T. meridian passage at Greenwich 21d.17h.44m.

daily difference 56m.

proportion for longitude $= 56 \times \frac{108^\circ}{360^\circ}$ = 17m. (to nearest minute)

L.M.T. meridian passage at Greenwich 21d.17h.44m.
longitude correction − 17m.

L.M.T. meridian passage at 108°E. 21d.17h.27m.
longitude in time − 7h.12m.

G.M.T. meridian passage at 108°E. 21d.10h.15m.
Zone (-7) + 7h.00m.

Zone time meridian passage at 108°E. 21d.17h.15m.

MERIDIAN PASSAGE OF A PLANET

The **L.M.T.**'s of the upper meridian passage of the four navigational planets are tabulated in the Abridged Nautical Almanac for every third day. Since the daily difference changes slowly, it is easy to obtain the time of transit for an intermediate date by inspection. If extreme accuracy is required and the daily difference is large, then a proportion must be applied to the time in order to obtain **L.M.T.** at the observer's meridian, as is done in the case of the moon.

EXAMPLE 44

Find the G.M.T. of the upper passage of Jupiter on the 9th January 1958 at an observer in longitude 120°W.

From the Abridged Nautical Almanac: Jupiter. upper passage Jan.
8 06.42
11 06.31
14 06.21

By inspection: L.M.T. meridian passage at Greenwich 9d.06h.38m.

long.correction $(3\frac{2}{3} \times \frac{120^\circ}{360^\circ})$ − 1m.

L.M.T. meridian passage at 120°W. 9d.06h.37m.

longitude in time + 8h.00m.

G.M.T. meridian passage at 120°W. 9d.14h.37m.

The meridian passage of a star is explained on page 59.

EXAMPLE 45

The observed meridian altitude of Betelgeuse(dec. 7°24'N) D.R.long 15°20'E was 25°10'.6 bearing N. Height of eye 40'. Find the latitude.

observed altitude	25° 10'.6 N.
dip	− 6'.1
apparent altitude	25° 04'.5
correction	− 2'.1
true altitude	25° 02'.4 N. (NX)
	90°
zenith distance	64° 57'.6 S. (ZX)
declination	7° 24'.0 N. (QX)
latitude	57° 33'.6 S. (QZ)

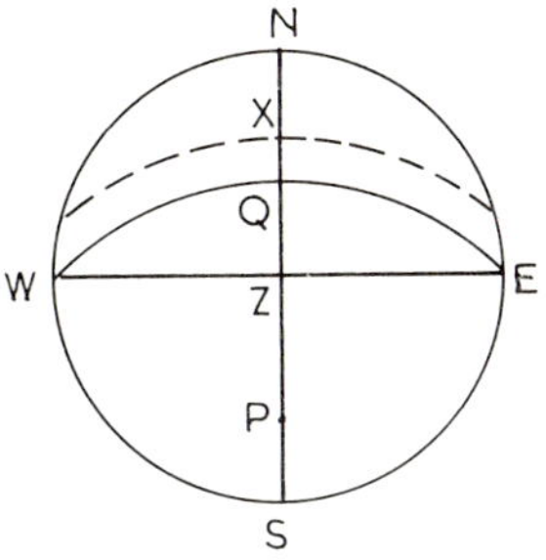

position line 090° - 270°

EXAMPLE 46

The true altitude of the star Kochab (dec. 74°20'0N.) DR. long. 20°30'W. on the meridian below the pole was 34°50'.5. Find the latitude of the observer.

Note: For a celestial body to be visible on the lower passage, the sum of the observer's latitude and the body's declination must be greater than 90°, with latitude and declination having the same name.

declination	74°20'.0 N.	(QX)
	90°	
polar distance	15°40'.0	(PX)
true altitude	34°50'.5	(NX_1)
latitude	50°30'.5 N.	(QZ)

position line 090° - 270°

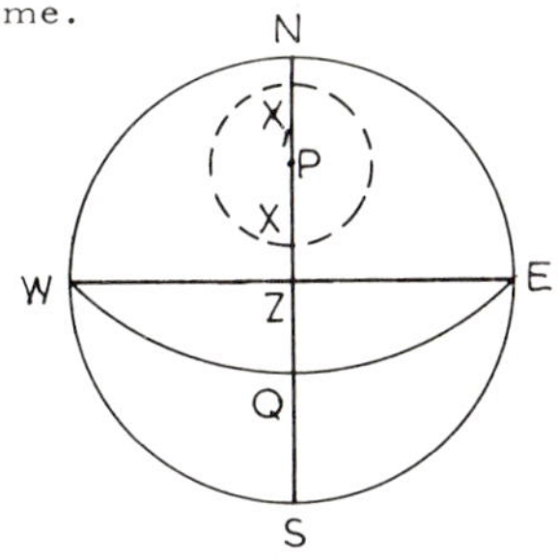

EXAMPLE 47

Find the sextant setting for the estimated meridian altitude of Fomalhaut (dec. 29° 52'.2 S.) for an observer in D.R. latitude 35°50'.0 N. Height of eye 50 feet. Index error 3'.0 off the arc.

latitude	35° 50'.0 N.	
declination	29° 52'.2 S.	
zenith distance	65° 42'.2 N.	(since star must pass south)
	90°	
true altitude	24° 17'.8 S.	
total correction	+ 2'.1	
apparent altitude	24° 19'.9	
dip	+ 6'.9	
observed altitude	24° 26'.8	
index error	− 3'.0	
sextant altitude	24° 23'.8 bearing of star 180°	

EXAMPLE 48

On the 20th October 1958 in longitude $12^{\circ}41'.0$E. the observed altitude of the Moon's upper limb was $36^{\circ}25'.0$ zenith N. H.E. 50 feet. Find the latitude and direction of the P.L.

L.M.T. (Greenwich) upper passage	20d.18h.54m.
correction for longitude	- 2m.
L.M.T. (ship) upper passage	20d.18h.52m.
longitude E.	- 50.7m.
G.M.T. upper passage	20d.18h.01.3m.

H.P. $56'.0$

declination (18h).		12° $43'.2$ S.
(d)	-	$0'.2$
declination		12° $43'.0$ S.

observed alt.	36° $25'.0$
dip	- $6'.9$
app.alt.	36° $18'.1$
correction	+ $58'.6$
upp.limb corr.	- $30'.0$
true altitude	36° $46'.7$
zenith dist.	53° $13'.3$N.
declination	12° $43'.0$S.
latitude	40° $30'.3$N.

P.L. 090° - 270°

EXAMPLE 49

23rd September 1958, in long: $20^{\circ}30'$E the sextant altitude of the Sun's upper limb on the meridian was $53^{\circ}11'.0$ bearing N. Index error $3'.2$ on the arc. Height of eye 29 feet. Find the latitude and direction of the position line.

Mer. passage	23d.11h.53m.
Long. E.	1h.22m.
G.M.T. mer.pass.	23d.10h.31m.

Declination	(10h)	00° $03'.1$N
increment	(31m)	$0'.5$
Declination		00° $02'.6$N

sex.alt.	53° $11'.0$ N.
ind.error	- $3'.2$
Obs.alt.	53° $07'.8$
dip.	- $5'.2$
app.alt.	53° $02'.6$
correction	- $17'.7$
true alt.	52° $44'.9$ N.
zenith dist.	37° $15'.1$ S.
declination	00° $02'.6$ N.
latitude	37° $12'.5$ S.

P.L. 090° - 270°

EXAMPLE 50

The observed meridian altitude of Mars was 49°40'.0 bearing N on the 19th June 1958, in longitude 45°W. H.E. 50 feet. Find the latitude and direction of P.L.

(Note - The longitude correction in this case is less than 1 minute)

L.M.T. (Greenwich) mer.passage.	19d.06h.47m.	dec. (9 hrs.)	01°21'.9N
longitude W.	+ 3h.00m.	d.correction +	0'.6
G.M.T. mer.passage	19d.09h.47m.	declination	01°22'.5N

observed alt.	49°40'.0N
dip	- 6'.9
app.altitude	49°33'.1
correction	- 0'.8
true alt.	49°32'.3N
zenith dist.	40°27'.7S
declination	01°22'.5N
latitude	39°05'.2S

P.L. 090° - 270°

LATITUDE BY POLE STAR

The proximity of the Pole Star (Polaris) to the celestial pole enables an observer to obtain the latitude of the point where the position circle cuts the meridian of the estimated position.

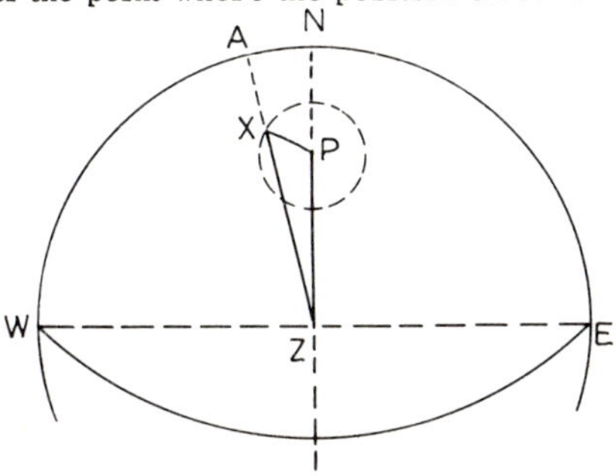

In the figure X is the position of Polaris with altitude AX. Since the altitude of Polaris cannot differ from the altitude of the elevated pole by more than about 1°, its altitude will differ from the observer's latitude by the same amount.

Therefore latitude = true altitude - a small correction.

This small correction is tabulated in the Abridged Nautical Alamanc under the heading 'Pole Star Correction Tables'.

There are three parts to the correction:- a_0, a_1, a_2.

To a_0 is added a constant 58'.8 to keep it positive,
to a_1 is added a constant 0'.6 to keep it positive,
and to a_2 is added a constant 0'.6 to keep it positive.

The sum of these constants is 60' and must be subtracted from the altitude before the latitude is obtained.

Therefore, latitude = true altitude of Polaris $- 1^\circ + a_0 + a_1 + a_2$

The Nautical Almanac is entered with the L. H. A. ♈ to avoid the necessity of finding the S. H. A. of Polaris which changes rapidly due to the precession of the celestial poles.

A table from which the azimuth of the Pole Star may be found is included in the Nautical Almanac.

EXAMPLE 51

On 10th October 1958 in longitude 39° 54'E. the sextant altitude of Polaris was 43° 12'.0, index error 3'.0 off the arc, when a chronometer correct for G.M.T. showed 17h.15m.27s. Height of eye 30 feet. Find the latitude and direction of the P.L.

G.M.T. 10d.17h.15m.27s.	G.H.A. ♈	273°47'.4
	inc.	3°52'.4
	G.H.A. ♈	277°39'.8
	Long. E.	39°54'.0
	L.H.A. ♈	317°33'.8

Sextant alt.		43°12'.0
index error		+ 3'.0
observed alt.		43°15'.0
dip.		- 5'.3
apparent alt.		43°09'.7
star correction		- 1'.0
	a_0	+ 41'.4
	a_1	+ 0'.5
	a_2	+ 0'.9
		-1°00'.0
	latitude	42°51'.5 N.

Az. of Pole Star 001°.3
P.L. 091°.3 - 271°.3

The problem of finding a position through which the circle of equal altitude passes is simple when the body is on the observer's meridian. When the body is not on the meridian, the problem becomes that of solving a spherical triangle.

There are three methods of obtaining the position line in the above case :-

1. Longitude by chronometer, when the body is close to the prime vertical, i.e.. the azimuth is close to due east or west.

2. Intercept method for any azimuth. This method is gaining popularity with the more progressive navigator.

3. Latitude by ex-meridian, which is restricted to bodies close to the observer's meridian.

THE LONGITUDE METHOD

The figure is drawn on the plane of the rational horizon. The latitude of the observer (QZ) is the latitude of the estimated position. It is required to find the L. H. A. of the observed body (P).

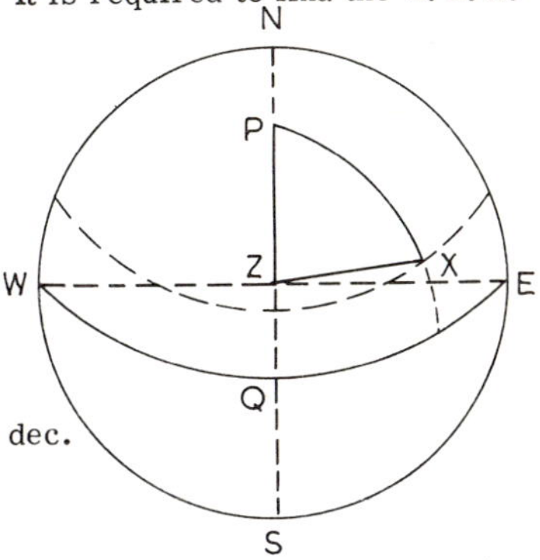

From the haversine formula:

$$\text{hav } P = \frac{\text{hav } ZX - \text{hav } (PZ \sim PX)}{\sin PZ.\sin PX}$$

which may be modified to:

$$\text{hav } P = \frac{\text{hav } ZX - \text{hav } (\text{lat} \pm \text{dec})}{\cos \text{lat.} \cos \text{dec.}}$$

finally

$$\text{hav } P = \left[\text{hav } ZX - \text{hav } (\text{lat} \pm \text{dec})\right] \sec \text{lat.} \sec \text{dec.}$$

Once the L. H. A. has been determined, the longitude may be found by comparison with the G. H. A. obtained from the Nautical Almanac.

The resultant longitude obtained by this method is the longitude on the parallel of estimated latitude through which the position line passes. A perpendicular line to the azimuth then gives the direction of the position line.

The following are examples in the use of the longitude method.

EXAMPLE 52

On January 10th 1958 p.m. at ship, in D.R. pos lat 11°00'N., long. 109°30'.0 E the observed altitude of the Sun's lower limb was 50°27'.0, height of eye 45 feet. The time by a chronometer correct for G.M.T. was 06h.17m.30s. Find the direction of the position line and the position through which to draw it.

G.M.T. 10d.06h.17m.30s.

$$\text{hav } P = \frac{\text{hav} ZX - \text{hav } (\text{lat} \pm \text{dec})}{\cos \text{lat.} \cos \text{dec.}}$$

observed alt.	50°27'.0	declination (06)	22°01'.5 S	nat.hav. ZX	0.11364
dip	- 6'.5	d	- 0'.1	nat.hav.(l ± d)	0.08078
apparent alt.	50°20'.5	declination	22°01'.4 S		
correction	+ 15'.4	latitude	11°00'.0 N	nat.hav.	0.03286
true alt.	50°35'.9	difference	33°01'.4	log.hav.	8.51665
zenith dist.	39°24'.1			log.sec.lat.	0.00805
				log.sec.dec.	0.03291
G.H.A. Sun (06)	268°09'.2			log.hav.P.	8.55761
increment +	4°22'.5				
G.H.A. Sun	272°31'.7 (GX)			∴ P = 21°54'.7W	
L.H.A. Sun	21°54'.7 (OX)				
	250°37'.0				
	360°00'.0				
longitude	109°23'.0E (GO)				

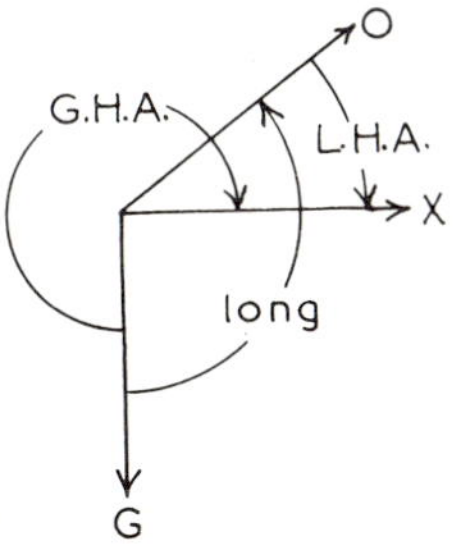

From A.B.C. tables		
	A = 0.48 S	
	B = 1.08 S	
	C = 1.56 S	∴ Azimuth = 213°.2

Answer: P.L. 123°.2 - 303°.2 runs through lat. 11°00'.0N. long. 109°23'.0E

EXAMPLE **53**

January 4th 1958 in D.R. position lat. 52° 03'N., long 20° 20'W. the sextant altitude of the planet Mars east of meridian was 11° 10'.5. Index error 1'.5 on the arc. Height of eye 32 feet. The time by a chronometer was 8h.42m.56s. Chronometer error 2m.01s. slow. Find the direction of the position line and the position through which to draw it.

$$\text{hav.P} = \left[\text{hav ZX} - \text{hav}\,(l \pm d)\right]\text{sec.lat.sec.dec.}$$

Chronometer		4d.08h.42m.56s.
error	+	2m.01s.
G.M.T.		4d.08h.44m.57s.

sextant alt.		11° 10'.5
index error		1'.5
observed alt.		11° 09'.0
dip	-	5'.5
apparent alt.		11° 03'.5
correction	-	4'.7
true alt.		10° 58'.8
zenith distance		79° 01'.2 (ZX)

declination (08)		21° 41'.1 S
d	+	0'.2
declination		21° 41'.3 S
latitude		52° 03'.0 N
difference		73° 44'.3

nat.hav. ZX	0.40476
nat.hav. (l ± d)	0.35999
nat.hav.	0.04477
log.hav.	8.65099
log.sec.lat.	0.21114
log.sec.dec.	0.03189
log.hav.P.	8.89402

∴ P = 327° 29'.5

G.H.A. Mars (08)	336° 38'.2
increment	11° 14'.3
v	0'.4
G.H.A. Mars	347° 52'.9
L.H.A. Mars	327° 29'.5
Longitude	20° 23'.4W

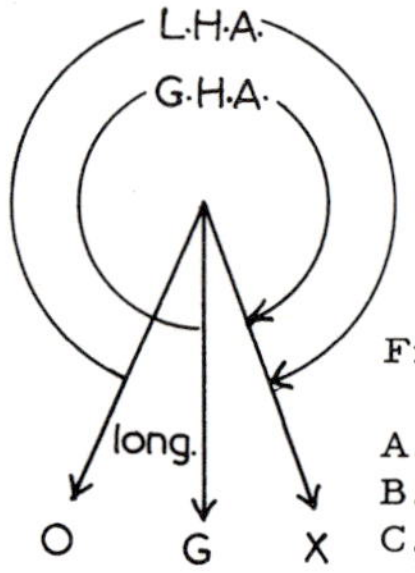

From A.B.C. tables:-

A.	2.01 S
B.	0.74 S
C.	2.75 S

∴ Az = 149°.4

Answer:- P.L. runs 059°.4 - 239°.4 through lat. 52° 03'N. long. 20° 23'.4W

THE INTERCEPT OR MARCQ ST. HILAIRE METHOD

The main point about this method is that it gives a comparison between the observer's known distance from the geographical position of the celestial body (the true zenith distance) and the calculated distance (calculated zenith distance) of the observer's estimated position from the same geographical position. The difference between the two is known as the intercept.

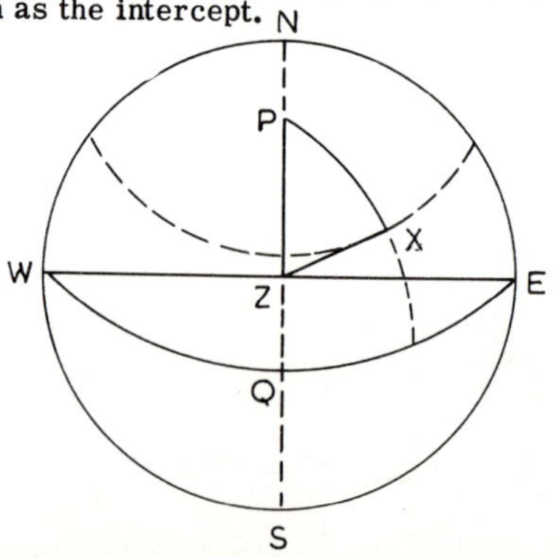

In this case, the part of the PZX triangle to be found is ZX.

From the haversine formula: $\text{hav } P = \dfrac{\text{hav } ZX - \text{hav }(PZ \sim PX)}{\sin PZ.\sin PX}$

re-arranging: $\text{hav } ZX = \text{hav } P.\sin PZ.\sin PX + \text{hav }(PZ \sim PX)$

which may be modified to: $\text{hav } ZX = \text{hav } P.\cos \text{lat}.\cos \text{dec} + \text{hav }(\text{lat} \pm \text{dec})$

Having determined the calculated zenith distance (C. Z. D.) it must be compared with the true zenith distance (T. Z. D.) to see whether the intercept is towards or away from the estimated position along the line of the azimuth.

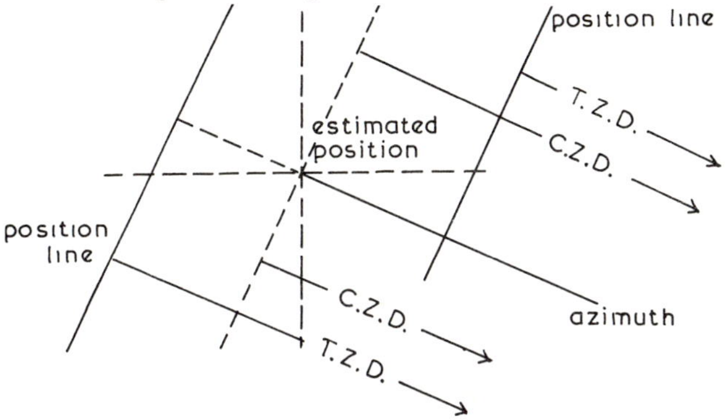

From the figure it can be seen that if the T. Z. D. is greater, then the intercept is away from the estimated position. This may be summarised as follows:

True greatest, intercept away,

True least, intercept towards.

The following are examples in the use of the intercept method.

EXAMPLE **54**

On 12th November in D.R. position lat. 40°04!.S., long. 100°00'W on a W'ly brg. the sextant altitude of the star Canopus was 39°24!.5 index error 2'0 off the arc. Height of eye 37 feet. The time by a chronometer correct for G.M.T. was 14h. 37m.28s. Find the direction of the position line and the position through which to draw it.

G.M.T. 12d.14h.37m.28s.

hav.ZX = hav.P coslat.cosdec + hav (lat ± dec.)

G.H.A. ♈ (14)	261°11!.5			log.hav.L.H.A.	9.56708
increment	9°23!.5			log.cos.lat.	9.88383
				log.cos.dec.	9.78275
G.H.A. ♈	270°35!.0			log.hav,	9.23366
S.H.A. Canopus	264°14!.0	declination	52°40!.3 S	nat.hav.	0.17126
		latitude	40°04!.0 S	nat.hav. (l ± d)	0.01205
G.H.A. Canopus	534°49!.0	difference	12°36!.3	nat.hav.ZX	0.18331
Longitude W	100°00!.0				
				ZX =	50°42!.0
L.H.A. Canopus	434°49!.0	from A.B.C. tables			
	74°49!.0W				

A =	0.22N
B =	1.36S
C =	1.14S
Az =	228°.9

Sextant altitude	39°24!.5
index error	+ 2!.0
observed altitude	39°26!.5
dip	- 5!.9
apparent altitude	39°20!.6
correction	- 1!.2
true altitude	39°19!.4
true zenith dist.	50°40!.6
calc.zenith dist.	50°42!.0
intercept. towards	1!.4

D.R. position	latitude 40°04!.0S.	long. 100°00!.0W
intercept	0!.9S	1!.4W
int.terminal pt. lat.	40°04!.9S	long. 100°01!.4W

P.L. 138°.9 - 318°.9

EXAMPLE 55

September 20th 1958 in D.R. position lat. 40° 12'.0S. long. 23° 48'W. the observed altitude of the Moon's upper limb was 63° 22'0 on a NW'ly bearing. Height of eye 50'. The time by a chronometer without error was 9h.15m.45s. Find the direction of the position line and a position through which to draw it.

Note:- Since Moon is to the west of the meridian the L.M.T. must be later than meridian passage.

Approx. L.M.T. mer.pass.		20d.18h.25m.
long.W.	+	1h.34m.
Approx. G.M.T. mer.pass.		20d.19h.59m.

hav ZX = hav P.cos l. cos d. + hav (l ± d)

G.M.T. 20d.21h.15m.45s.

G.H.A. Moon (21h)	37°19'.9	declination (21)	18°14'.3 S
increment	3°45'.5	d	- 0'.6
v	2'.2	declination	18°13'.7 S
G.H.A. Moon	41°07'.6	latitude	40°12'.0 S
longitude	23°48'.0	difference	21°58'.3
L.H.A. Moon	17°19'.6 W	H.P. 57'.4	

observed altitude	63°22'.0	log.hav.L.H.A.	8.35581
dip	- 6'.9	log.cos.lat.	9.88298
apparent altitude	63°15'.1	log.cos.dec.	9.97764
correction	+ 39'.7	log.hav.	8.21643
correction for U.L.	- 30'.0	nat.hav.	0.01646
true altitude	63°24'.8	nat.hav. (l ± d)	0.03631
true zenith dist.	26°35'.2	nat.hav.ZX	0.05277
calc.zenith dist.	26°33'.7		
intercept away	1'.5	∴ ZX = 26°33'.7	

from A.B.C. tables

A = 2.70N	D.R.lat.	40°12'.0 S	long 23°48'.0W
B = 1.11S	Intercept	1'.2 S	1'.2E
C = 1.59N	terminal pos.	40°13'.3 S	23°46'.6W

Az = 320°.5

P.L. 050°.5 - 230°.5

THE EX-MERIDIAN METHOD

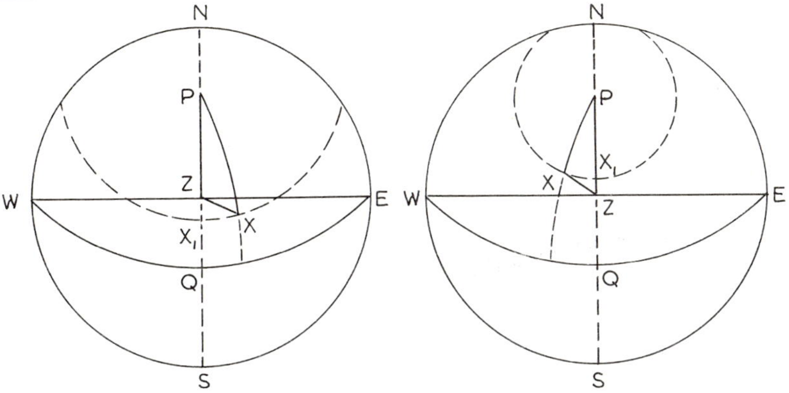

In the figures, X is a celestial body so close to the meridian that the change of declination while moving from X to X_1 is virtually negligible.

Then:- $PX_1 = PX$ and :- $ZX_1 = PX_1 \sim PZ$

therefore:- $M.Z.D. = PX \sim PZ$

By the haversine formula: $\text{hav } P = \dfrac{\text{hav } ZX - \text{hav } (PZ \sim PX)}{\sin PZ . \sin PX}$

re-arranging: $\text{hav } (PZ \sim PX) = \text{hav } ZX - \text{hav } P . \sin PZ . \sin PX$

or: $\text{hav } M.Z.D. = \text{hav } ZX - \text{hav } P . \sin PZ . \sin PX$

which may be modified to: $\text{hav } M.Z.D. = \text{hav } ZX - \text{hav } P . \cos \text{lat} . \cos \text{dec}.$

The latitude is then obtained by applying the declination of the observed body to the M. Z. D.

It is important to remember that the latitude so found is the latitude at the time of observation on the estimated meridian through which the position line passes in the direction perpendicular to the azimuth of the body's geographical position.

Note: When using the ex-meridian formula , the difference between latitude and declination should never be less than 4°, and the L. H. A. of the observed body, measured in minutes of time, should not be more than the zenith distance of the body in degrees.

An alternative approach lies in the short method ex-meridian tables published in both Norie's and Burton's tables.

Table I is entered with D.R. latitude and declination and a factor "A" (in Norie's) "F" (in Burton's) is obtained. This quantity is the change in altitude for one minute.

Table II is entered with "A" or "F" depending which tables are being used, and the L. H. A. and the reduction to the true altitude is obtained. It is plus at upper transit and minus at lower transit.

Table III contains an additional correction to the "reduction" which must be used when the "reduction" is large.

Table IV is an auxiliary table for determining the limits of the L. H. A. for ex-meridian problems.

The following are examples in the use of the ex-meridian formula.

EXAMPLE 56

On 14th January 1958 in D.R. position lat. 27°02'N., long 16°15'W. the observed altitude of the Sun's lower limb, near the meridian was 41°20'.4. Height of eye 46 feet. The time by a chronometer which was 15m.20s. slow on G.M.T. was 1h.13m.44s. Find the direction of the position line and a position through which to draw it.

$$\text{hav}\,(l \pm d) = \text{hav ZX} - \text{hav.P.coslat.cosdec.}$$

from N.A.	L.M.T. mer.passage	12.09
	longitude	1.05
	G.M.T. mer.passage	13.14

Chronometer	14d.13h.13m.44s.
error	+ 15m.20s.
G.M.T.	14d.13h.29m.04s.

G.H.A. sun (13h)	12°44'.3
increment	7°16'.0
G.H.A. sun	20°00'.3
longitude W	16°15'.0
L.H.A. sun	03°45'.3W

declination (13h)		21°20'.0S
d	-	0'.2
declination		21°19'.8S

observed altitude	41°20'.4
dip	6'.6
apparent altitude	41°13'.8
correction	15'.2
true altitude	41°29'.0
zenith distance	48°31'.0

log.hav.L.H.A.	7.03086
log.cos.lat.	9.94975
log.cos.dec.	9.96918
log.hav.	6.94979
nat.hav.	0.00089
nat.hav.ZX	0.16880
nat.hav.(l ± d)	0.16791
=	48°22'.8

lat. ± dec.	48°22'.8
dec.	21°19'.8 S
latitude	27°03'.0 N

from A.B.C.	A.7.78 S
	B.5.98 S
	C.13.76 S
	= Az 184°.6

Position line 094°.6 274°.6 through lat. 27°03'.0N., long 16°15'.0W.

alternatively

latitude 27°02'.0N. declination 21°19'.8S. L.H.A. 03°45'.3

"A" = 2.17 (it is important to interpolate accurately)

from Table II reduction =	8'.14
true altitude	41°29'.0
	41°37'.14
M.zenith dist.	48°22'.86
declination	21°19'.80 S
latitude	27°03'.06 N

The azimuth must be obtained by the ABC tables as before.

At the risk of appearing repetitious it must be emphasised that the latitude obtained is not the latitude of the Ship but the latitude on the selected meridian through which the P.L. passes.

EXAMPLE 57

On 15th June in D.R. position latitude 31°20'S., longitude 155°40'E the sextant altitude of the star Denebola was 43°00'.7. Index error 2.0 mins. on the arc. Height of eye 35 feet. The time by a chronometer correct for G.M.T. was 07h.12m.13s. Find the direction of the position line and the latitude through which to draw it.

$$\text{hav}\,(l \pm d) = \text{hav}\,ZX - \text{hav}\,P.\cos.\text{lat}.\cos.\text{dec}.$$

G.M.T. 15d.07h.12m.13s.

G.H.A. ♈	(07h)	008°03'.5
increment		3°03'.8
G.H.A. ♈		011°07'.3
S.H.A. Denebola		183°15'.8
G.H.A. Denebola		194°23'.1
longitude E.		155°40'.0
L.H.A. Denebola		350°03'.1W = 009°56'.9E .

declination 14°48'.3N

sextant altitude	43°00'.7
index error	- 2'.0
observed altitude	42°58'.7
dip	- 5'.7
apparent altitude	42°53'.0
correction	- 1'.0
true altitude	42°52'.0
zenith distance	47°08'.0

log.hav.L.H.A.	7.87611
log.cos.lat.	9.93154
log.cos.dec.	9.98532
log.hav.	7.79297
nat.hav.	0.00621
nat.hav.ZX	0.15985
nat.hav. (l ± d)	0.15364
lat. ± dec.	46°09'.3
declination	14°48'.3N
latitude	31°21'.0S

from A.B.C. tables:-

A = 3.47N
B = 1.52N
C = 4.99N

Azimuth 013°.2

Position line runs 103°.2 - 283°.2 through latitude 31°21'.0S longitude 155°40'E.

Using ex-meridian tables:-

latitude 31°20'S L.H.A. 009°56'.9E
declination 14°48'.3N

from Table I "A" = 2.267

from Table	II	reduction	59'.84
from Table	III	2ndcorr.	0'.4
		reduction	59'.44
		true altitude	42°52'.0
			43°51'.44
		M.zen.dist.	46°08'.56
		declination	14°48'.3 N
		latitude	31°20'.26 S

Position line as in previous working

X
PLOTTING

POSITION LINES AND PLOTTING

A single position line obtained either from a single celestial observation or from the bearing of a terrestrial object gives no information regarding the ship's position, simply that the vessel is somewhere on that line.

To obtain the observed position of a vessel, the navigator must obtain more than one position line, either simultaneously or combined with a known run between observations.

ERRORS IN POSITION LINES

1. Error in longitude due to an error in latitude.

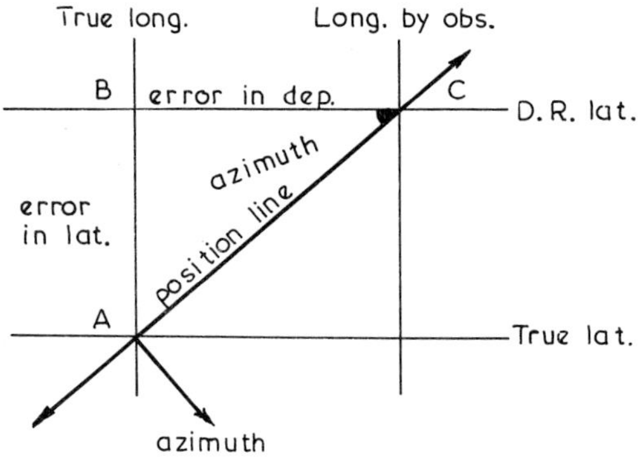

In the triangle ABC: AB = error in latitude
BC = error in departure from AB
angle BCA = azimuth
error in dep.= AB. cot (BCA)
dep.= dlong. cos (lat)

therefore: AB. cot(BCA)= dlong. cos (lat)

re-arranging: dlong = error in longitude = error in lat. cot az. sec lat.

It will be noticed that cot az. sec lat is quantity C in ABC tables which enables the longitude to be corrected quickly.

2. Error in position line due to error in altitude.

The next figure shows the result of error in the observed altitude. Where the observed altitude is too small, the position line is displaced away from the object, and where the observed altitude is too large, the position line is displaced towards the object. This could be caused by failing to apply the index error of the sextant, or applying the index error incorrectly. Errors in the true dip due to abnormal refraction will have the same effect.

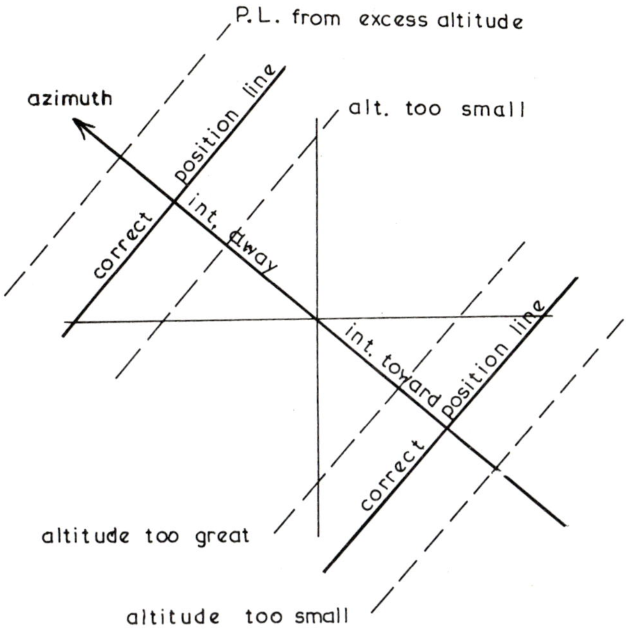

In the figure below

PL represents the correct position line passing through the true position A.

P_1L_1 represents the displaced position line due to the error in altitude AB.

AC = error in latitude due to AB

AD = error in departure due to AB

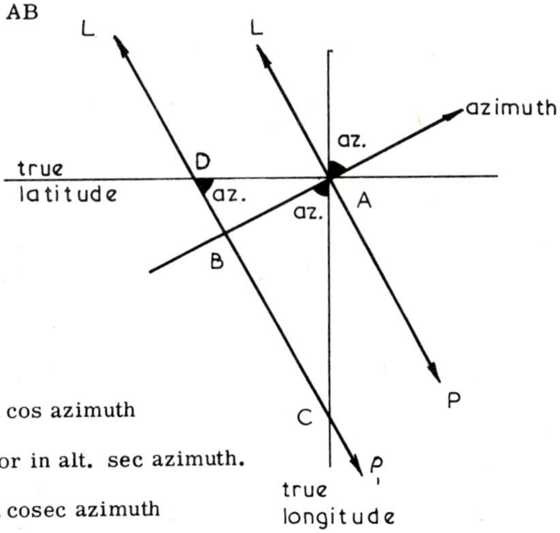

in triangle ABC AB = AC. cos azimuth

therefore: error in lat = error in alt. sec azimuth.

in triangle ABD AD = AB. cosec azimuth

or: error in dep. = error in alt. cosec az.

but: error in dep. = error in long. cos lat.

therefore: error in long = error in alt. sec lat. cosec az.

3. Error in position line due to error in time.

In the figure, PAL is the correct position line obtained with the correct hour angle, P_1BL_1 is the position line obtained with the false hour angle.

AB = error in zenith distance due to error in H. A.
BC = error in departure due to error in H. A.

angle DAB = azimuth
angle ACB = $(180^o$ - az.)

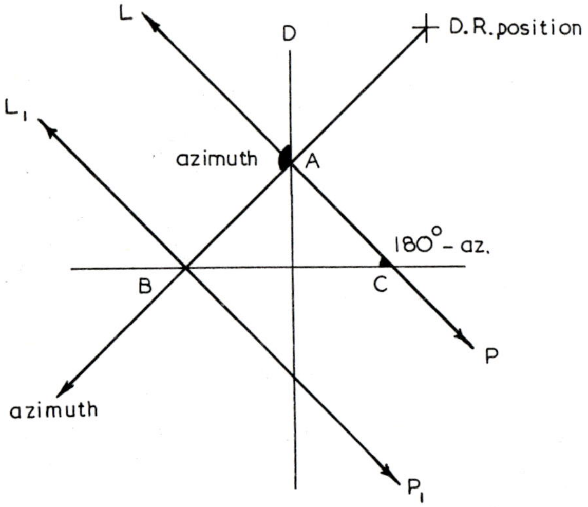

in triangle ABC: BC = AB. cosec $(180^o$ - az.)

or: error in dep. = error in zen. dist. cosec az. .... (1)

but: error in dep. = error in long. cos. lat. .... (2)

equating (1) and (2) and re-arranging:

error in zen. dist. = dlong. sin az. cos lat.

Application of this formula will show that the error in longitude (BC) is equal to the error in G. M. T., both expressed in minutes of arc.

If the G. M. T. is too large, the position line is shifted to the west.

If the G. M. T. is too small, the position line is shifted to the east.

EXAMPLE 58

An observer in D.R. position latitude 43^oN., longitude 47^oW., obtained an intercept of 2!5 towards with an azimuth of 136^oT. It was then found that the index error of the sextant of 2!0 had been added instead of subtracted and that the chronometer error of 17s. fast had also been applied the wrong way. Find the correct intercept.

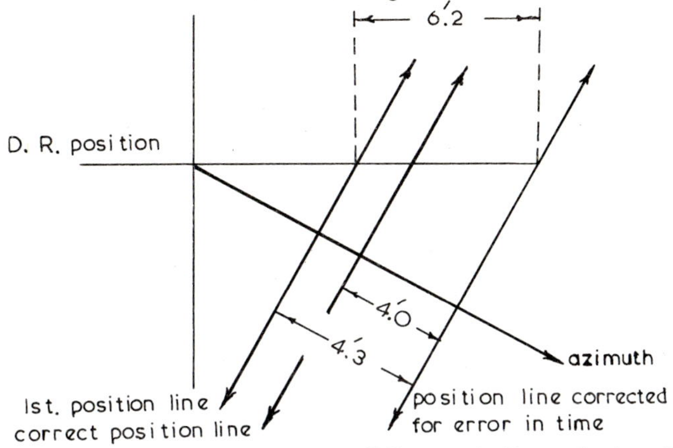

G.M.T. is 34s. too large = 8'.5 error in longitude therefore position line is 8'.5 of longitude too far to the west and must be moved 8'.5 of long, or 6'.2 of departure to the east.

error in dep. = error in zen. dist. cosec. az.

or: error in zen.dist. = error in dep. sin azimuth.
= 6.2. sin 136°
= 4.3 towards. (see figure)

error in altitude due to the wrongly applied index error = error in intercept = 4'.0 away. (see figure)

therefore the correct intercept is: 2'.5 + 4'.3 - 4'.0 = 2'.8 towards.

The most convenient method of ascertaining the observed position is to plot the results of the celestial observations on the Mercator chart. The intersection of the position lines denotes the observed position. However, the scale of the chart may be too small to give the required degree of accuracy. The same result can then be obtained by using a blank sheet of paper. Since no scale is provided it must be remembered that departure must be converted into longitude before the final position can be obtained.

EXAMPLE 59

In D.R. position latitude 35°08'S., longitude 26°42'W. an intercept of 3'.0 away was obtained from a star bearing 165°. After steering 271°T. for 25 miles an observation gave an intercept of 1'.3 towards bearing 070°. Find the position of the vessel at the second observation.

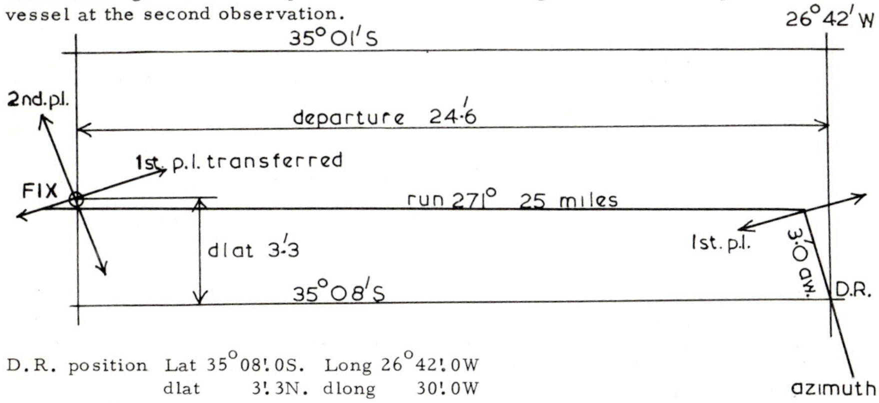

D.R. position Lat 35°08'.0S. Long 26°42'.0W
dlat 3'.3N. dlong 30'.0W

Obs.position Lat 35°04'.7S. Long 27°12'.0W

EXAMPLE 60

At 0900 a point of land in latitude 26°54'N., longitude 35°15'E. bore 327°T. The vessel then steered 035°T. at 15 knots. At 1200 an ex-meridian observation of the Sun bearing 174° gave a position line passing through latitude 27°34'N., longitude 35°52'E. The current setting throughout 310° at 2 knots. Find the vessel's position at 0900 and 1200.

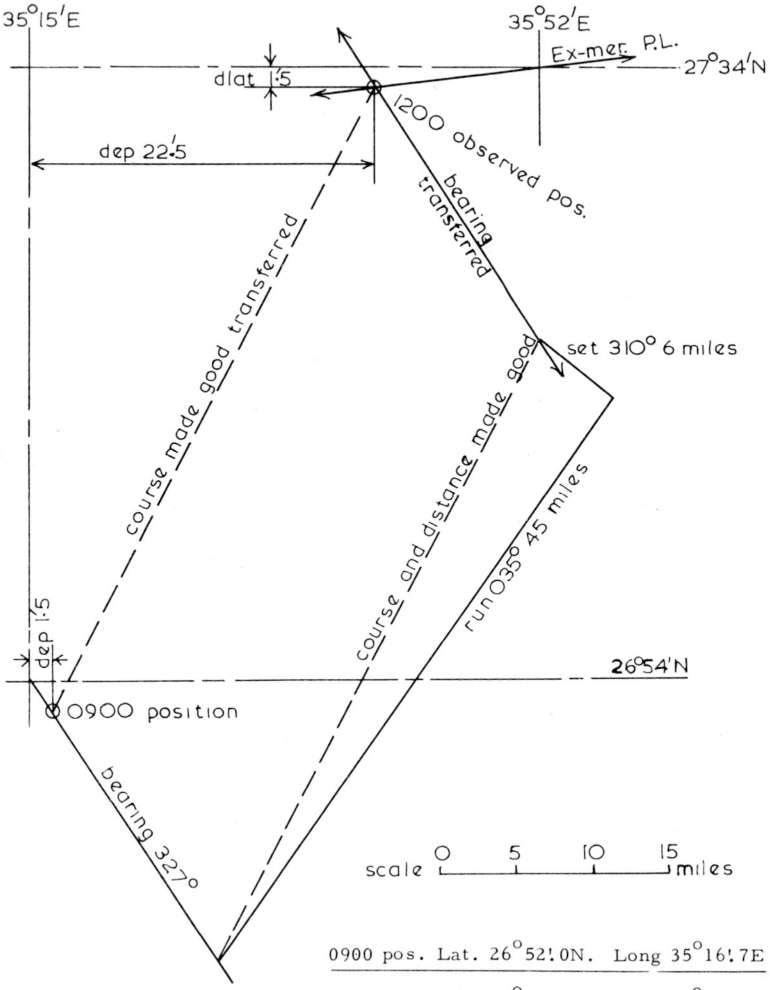

0900 pos. Lat. 26°52'.0N. Long 35°16'.7E

1200 pos. Lat. 27°32'.9N. Long 35°40'.0E

EXAMPLE 61

A vessel in position 25°34'N., longitude by observation 115°16'E. azimuth of Sun 113°, steered 220°T. at 16 knots. Three hours later a meridian altitude observation of the Sun gave a latitude of 25°02'N. By means of a plot obtain the vessel's position at the time of the meridian altitude observation.

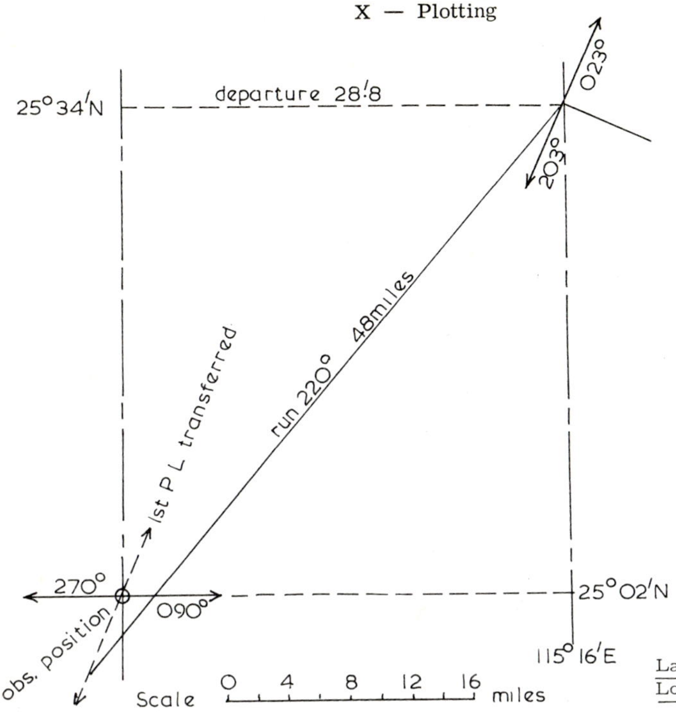

Obs. position

Latitude	25°02'.0N
Longitude	114°44'.1E

EXAMPLE **62**

In D.R. position latitude 37°20'N., longitude 42°01'W. the following observations were taken:

intercept 3'.8 towards with the star bearing 245°.
intercept 4'.6 away with the Moon bearing 146°: Plot the observed position.

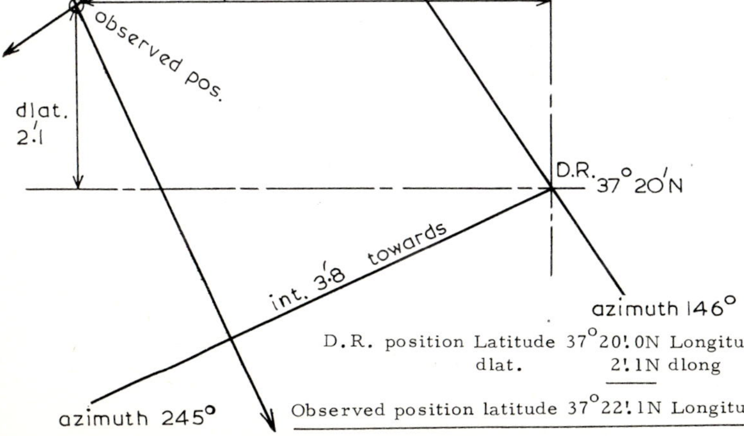

D.R. position Latitude	37°20'.0N	Longitude	42°01'.0W
dlat.	2'.1N	dlong	6'.5W
Observed position latitude	37°22'.1N	Longitude	42°07'.5W

NAVIGATIONAL FIGURE DRAWING

The most popular projection for navigational problems is the equidistant projection. It has the merit of convenience and simplicity and offers a good approximation for the desired answers.

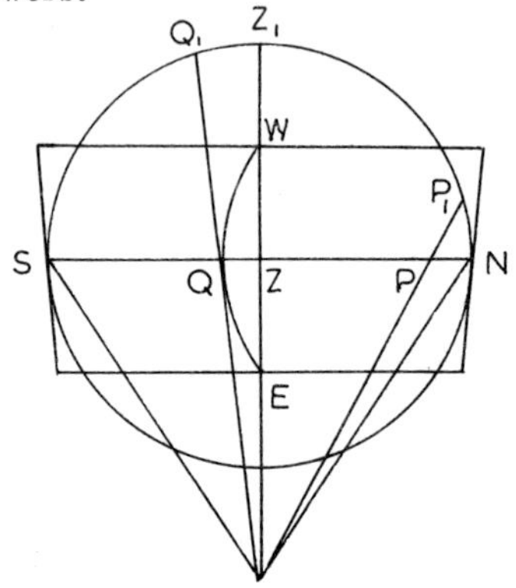

The eye is situated on a diameter extended beyond the sphere by $\sqrt{\frac{1}{2}}$ radius. All points are then projected on to a plane surface that is perpendicular to that diameter.

Method of Construction.

1. With centre Z (observer's zenith) construct a circle to some convenient radius. i.e. $\frac{9}{10}$ or $\frac{18}{10}$". This affords an easy scale for angular measure each tenth of an inch representing 10° or 5° respectively.

2. Construct the observer's meridian NZS, and the prime vertical WZE.

3. Insert the elevated pole P and also Q. (PZ = co-lat and ZQ = lat).

4. Construct the great circle WQE. This represents the equinoctial.

5. To construct the small circle of declination a'a a', first insert a (Qa = declination). Enter the amplitude tables to find Ea' and Wa' and lay these off by means of a protractor from centre Z. Finally construct arc a' a a'.

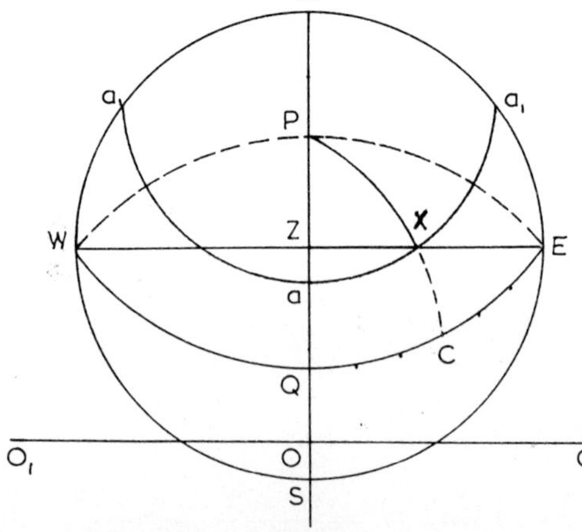

6. Obtain point 0 which is the centre of the circle passing through WPE by drawing the perpendicular bisector of the line WP towards NS.

7. Construct 0'0 0' perpendicular to PO. This is the locus of all hour circles or celestial meridians passing through P.

8. Divide the equinoctial into twelve equal parts and mark the appropriate hour angle.

9. Construct the hour circle PC and continue to X on the circle of declination. In this diagram the hour angle used is 3 hours east.

10. Complete the PZX triangle.

The azimuth is obtained by measuring angle PZX by protractor. The zenith distance by measuring ZX on the selected scale.

EXAMPLE 63

Given latitude 40°N. Declination of body 20°S. Hour angle 3hrs.30mins. west of meridian passage. Find the azimuth and the zenith distance of the body.

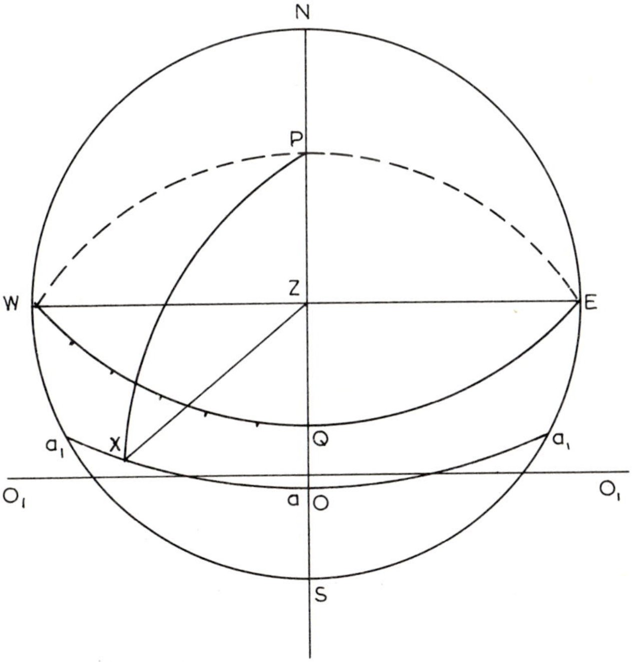

From the amplitude tables, using lat 40° and dec. 20° the true amplitude is East or West 26.5° South.

From the diagram the azimuth, angle PZX = 130° = 230° true azimuth.

zenith distance ZX = 77°

EXAMPLE 64

Draw a figure using the equidistant projection illustrating the great circle track from A in lat. 40°N., long 70°E., to B in lat 20°N, long 175°E. Show the composite track between these two places with limiting latitude 40°N.

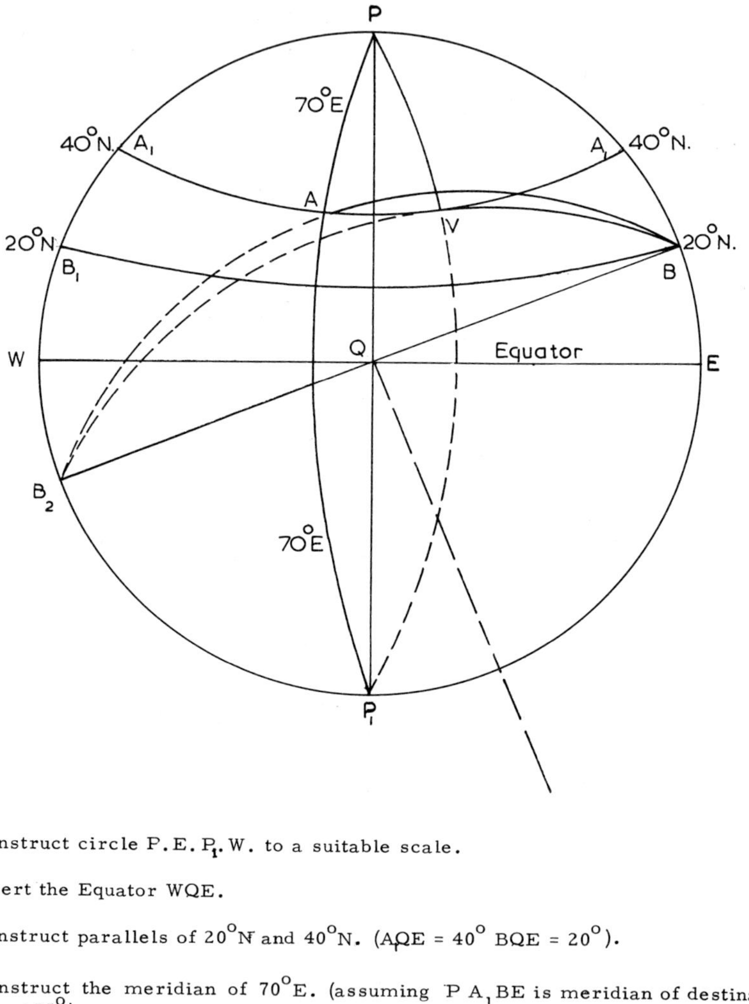

1. Construct circle P.E.P_1.W. to a suitable scale.

2. Insert the Equator WQE.

3. Construct parallels of 20°N and 40°N. (A_1QE = 40° BQE = 20°).

4. Construct the meridian of 70°E. (assuming P A_1BE is meridian of destination i.e. 175°).

5. The locus of the great circle track through A and B lies on the perpendicular bisector of BQB_2.

6. The composite track requires the construction of a great circle tangential to arc A_1A_1. The locus of this lies again on the perp. bisector of BQB_2. The vertex at the point of contact being V. The track consists of parallel sailing AV, and great circle sailing VB.

EXAMPLE 65

Latitude of observer 50°N. Declination of star 56°N. Find the altitude at upper and lower transits.

Show that: at lower transit latitude = altitude + polar distance

at upper transit latitude = zen.dist. ± declination

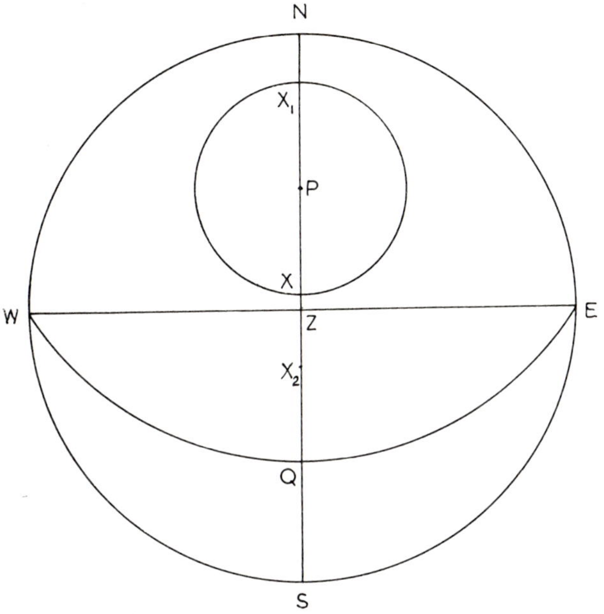

In this example, the figure is relatively simple. The star is visible on both upper and lower transits of the observer's meridian i.e. it is termed circumpolar.

The altitude at upper transit NX = 84°
The altitude at lower transit NX_1 = 16°

In the diagram PN = ZQ = latitude. (NZ = 90° and PZ is the co-lat then NZ - PZ = latitude = PN)

$$PN = NX_1 + PX_1$$

or, latitude = altitude at lower transit + polar distance.

In the diagram QZ = QX - ZX (if the body had passed to the south of the observer, i.e. X_2, this would have been:-

$$QZ = QX_2 + ZX_2)$$

therefore latitude = zenith distance ± declination.

EXAMPLE 66

Draw a figure using the equidistant projection, on the plane of the observer's meridian in latitude 40°S. Insert two stars:-

A with dec. 10°N on the meridian above the pole
B with dec. 70°S on the meridian below the pole

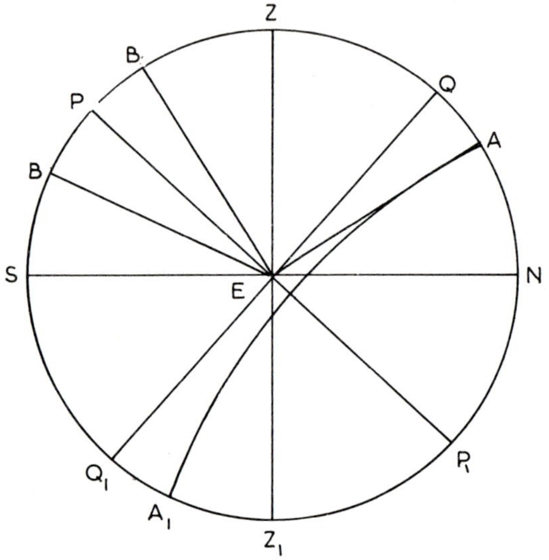

1. Construct the circle representing the plane of the observer's meridian to a suitable scale.

2. Insert ZEZ_1 the prime vertical where Z is the observer's zenith.

3. Construct angle ZEQ (40°) equal to the latitude and insert line QEQ_1 the equinoc tial.

4. Construct angle ZEP (50°) equal to the co-lat and insert the elevated pole P.

5. Construct angles QEA and QEB, equal to the declination of the two stars and complete the parallels of declination.

6. Construct angle PEB equal to PEB_1, the polar distance of star B (20°). Point B is the position of meridian passage of B below pole.

Angle NEA = true altitude of star A
" ZEA = true zenith distance of A

Angle SEB = true altitude of star B below the pole
" ZEB = true zenith distance of B

A more accurate projection is the stereographic projection. In this case the eye is situated on the surface of the sphere at the end of a diameter. All points are projected onto a plane surface perpendicular to that diameter.

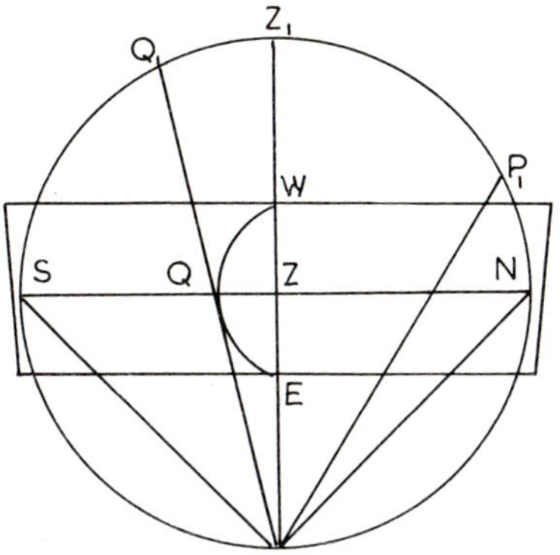

A scale circle must be constructed to the same dimensions that the figure is to be drawn. All measurements on the NZS and WZE lines must be taken from the diameter of the scale circle.

All angles extended to the NESW circle must be measured from the point diametrically opposite.

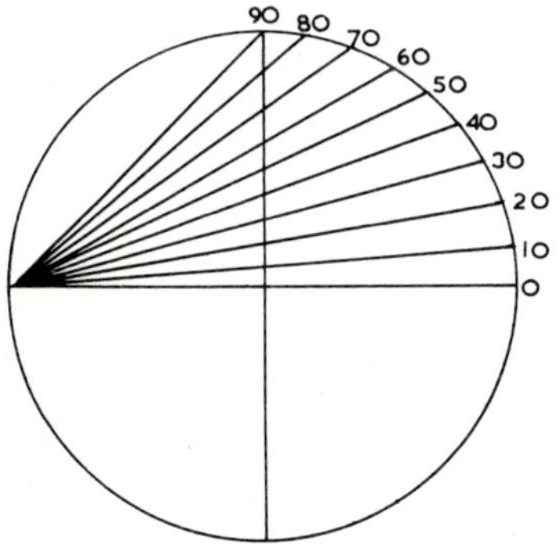

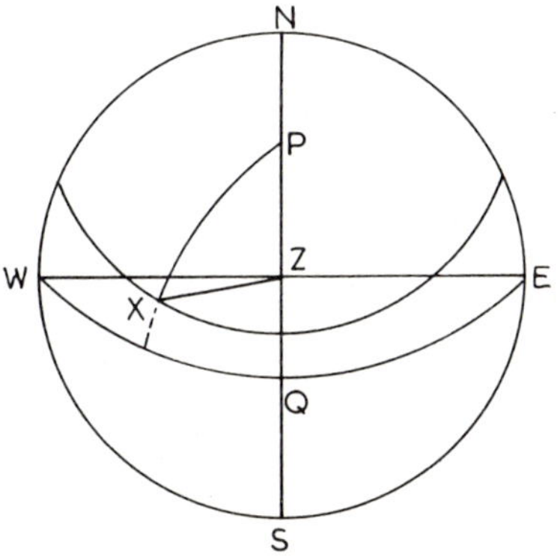

The stereographic projection gives accurate results, but on the whole takes more time than the more popular equidistant projection.

An example illustrating the use of the stereographic projection follows.

EXAMPLE **67**

Using the stereographic projection, construct a figure on the plane of the rational horizon in latitude 45°N. Show the parallel of declination of star X (20°N) and star Y (25°S). From the figure measure the zenith distances of the two stars in question on meridian passage.

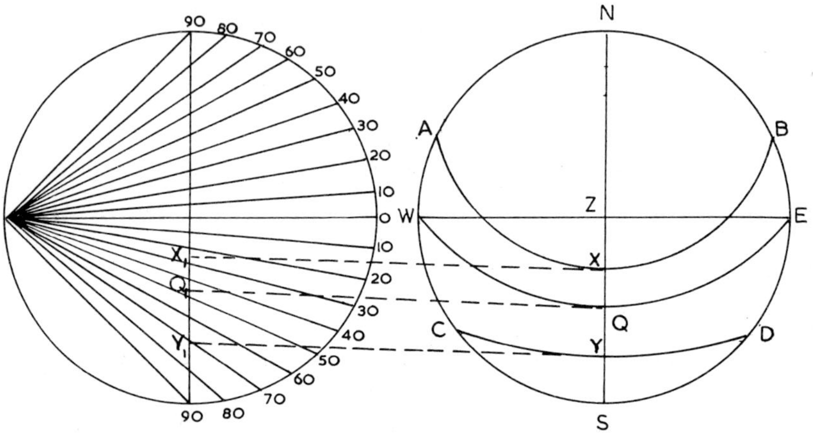

1. Draw scale circle and graduations

2. Construct circle NESW to same radius and on same level as scale circle.

3. Insert observer's meridian NZS and prime vertical WZE.

4. From diameter of scale circle insert Q 45° below Z.

5. Construct arc of circle WQE the equinoctial.

6. Insert QX (20°) and QY (25°) both measured along diameter of scale circle from Q.

7. From the amplitude tables obtain the angles WEA = EWB = 29° and WEC = EWD = 37°.

8. Construct the parallels of declination - AXB and CYD the locus of which lie on the NZS line extended.

To obtain zenith distance of X.

Measure ZX along centre line of scale circle = 25°.

To obtain zenith distance of Y.
Measure ZY along centre line of scale circle = 70°.

XI
TIDES

TIDES AND TIDAL PROBLEMS

The word 'tide' is often loosely used to describe both vertical and horizontal movements of the sea, so the following definitions must be adhered to:

TIDE - a periodical movement in the level of the sea surface, due to periodical forces.

TIDAL STREAMS - the horizontal movements of the sea that are due to periodical forces.

CURRENTS - horizontal movements of the sea, due to permanent, or temporary forces that are not periodical.

If observations are made of the changes in the sea level at any place, it will be seen that it is continually rising and falling in slightly irregular periods of time, closely resembling simple harmonic motion.

AMPLITUDE - this is the vertical displacement of high or low tide from some mean level.

DURATION - this is the time interval between successive high and low tides.

RANGE - this is the vertical distance between successive high and low tides.

CAUSES OF THE TIDES

Under ideal conditions the Earth is considered to have a uniform covering of water with no land masses. This diagram shows that if the Earth rotates within an ellipsoid of water formed by external forces, the point A will be carried round and the water level will periodically rise and fall.

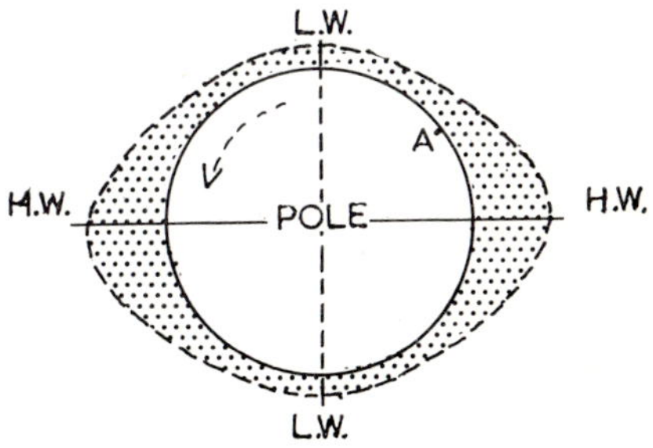

The main factor that must be considered is the attractive force between the Earth and its immediate planetary neighbour, the Moon. To a lesser extent, the Sun also exerts an influence upon the tides.

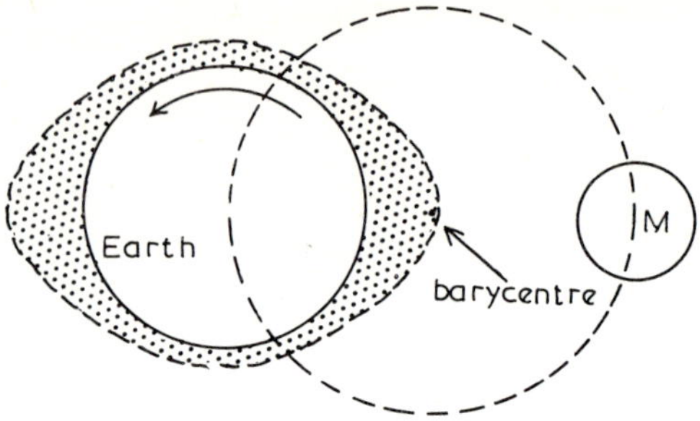

In the diagram, the Earth and Moon are shown revolving around their common centre of gravity, known as the barycentre. The attraction of the Moon causes the water to pile up towards the Moon. This also occurs, to a lesser extent, in the hemisphere remote from the Moon. This differential attraction causes an ellipsoid of water on the Earth's surface, and as the Earth rotates, the water level at any place would rise and fall in a lunar tide, with a period between successive high tides of half a lunar day, about 12 hours 25 minutes. The Sun also exerts a differential attraction on the water level, which is slightly less than half that of the Moon. This causes a solar tide with a period of twelve hours. The combined effects of Sun and Moon produce a luni-solar tide.

The remaining members of the Solar system are considered to have no effect on the tides.

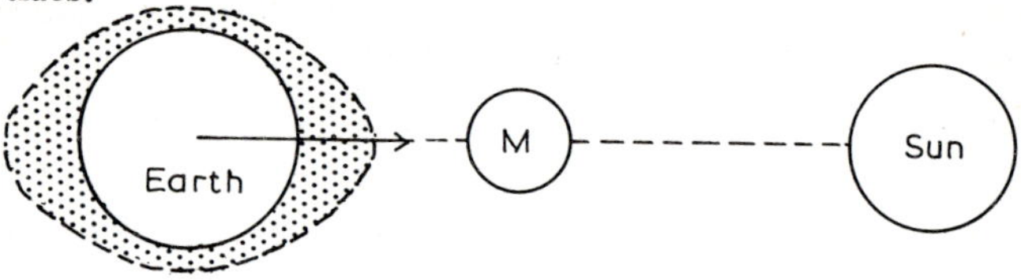

When the Sun and Moon are in conjunction, they will exert their attractive forces in the same direction, causing a much higher high tide.

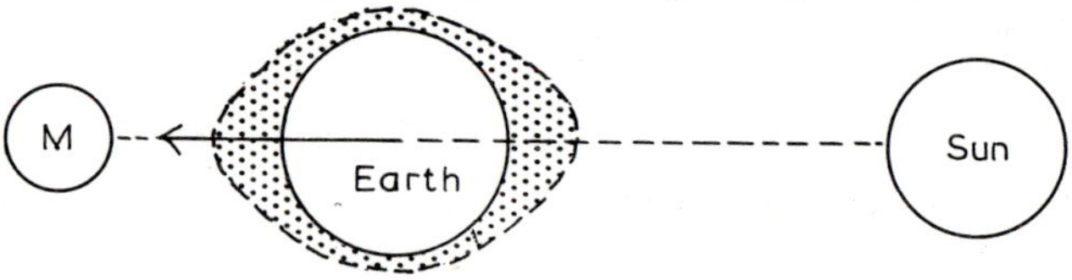

The same effect is produced when the Sun and Moon are in opposition. These maximum tides are known as Spring Tides.

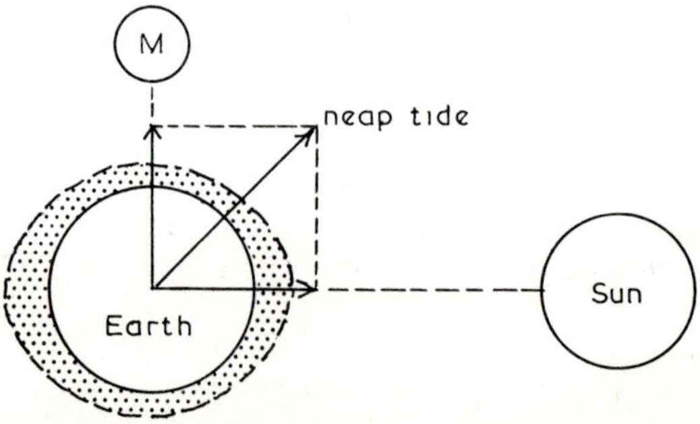

When the Sun and Moon are in quadrature, there is a much more uniform distribution of water, resulting in a lower high tide which is known as the Neap tide.

When the Moon is moving from its position in conjunction or opposition to that of quadrature, i. e. the first or third quarter the high tide will occur before the meridian passage of the moon and the tide is said to prime.

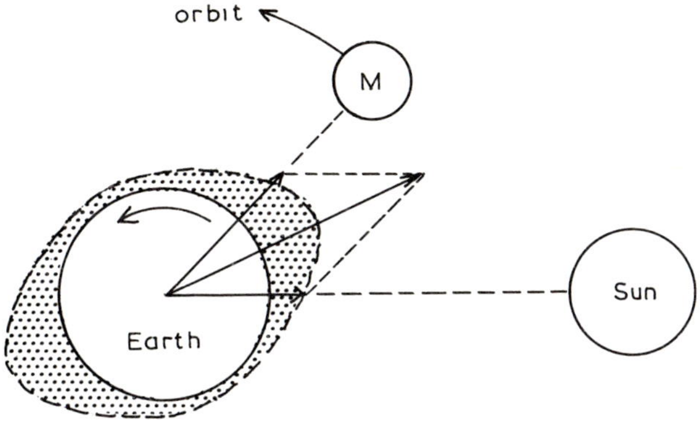

During the second and fourth quarters, when the Moon is moving away from quadrature the high tide will occur after the meridian passage of the Moon and the tide is said to lag.

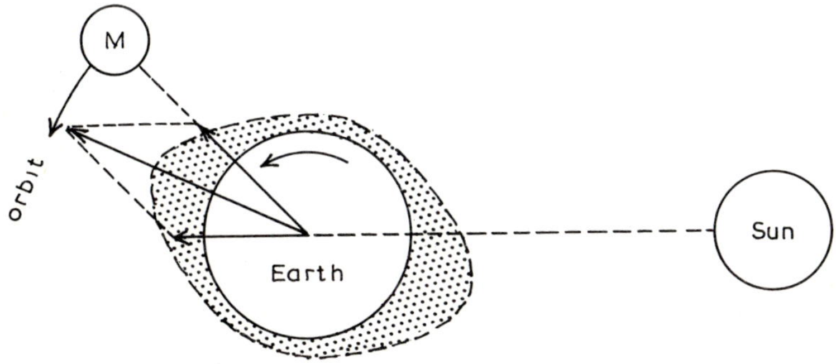

The conditions referred to, have been ideal. Nevertheless, actual observations agree very closely with the ideal conditions. Any differences that do occur, are due to a combination of several possible causes: the interception of the tidal wave by land masses, varying depths of the oceans, meteorological effects and the change in the relative positions of the Sun, Earth and Moon throughout the year.

The following diagram illustrates the various tidal levels:-

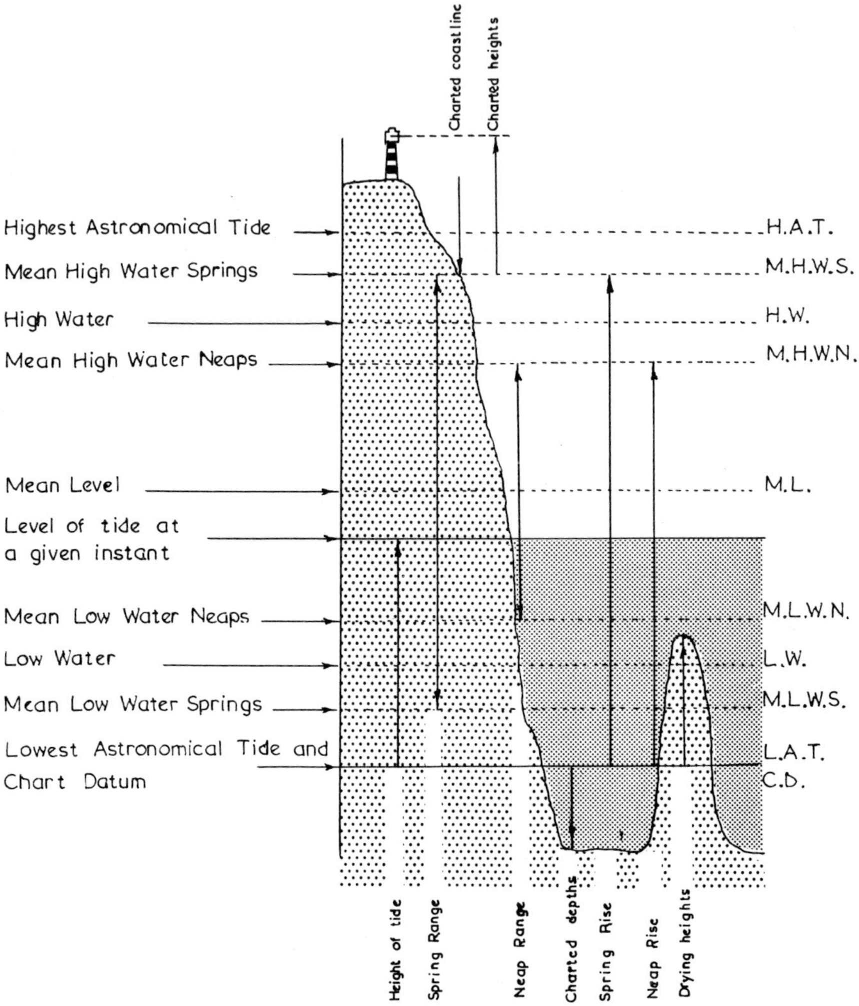

MEAN LOW WATER SPRINGS (M. L. W. S.). The average height, throughout a year when the average maximum declination of the moon is $23\frac{1}{2}^{o}$, of two successive low waters during those periods of twenty four hours (approx. once a fortnight) when the range of the tide is greatest.

MEAN HIGH WATER SPRINGS (M. H. W. S.). The average height of two successive high waters during the same periods.

MEAN LEVEL (M. L.). The mean of the heights M. H. W. S. , M. L. W. S. , M.H.W.N. , M. L. W. N.

MEAN HIGH WATER NEAPS (M. H. W. N.) AND MEAN LOW WATER NEAPS (M. L. W. N.). This is the average height, during the period described in M. L. W. S. of two successive high and low waters, respectively, when the range of the tide is least.

LOWEST ASTRONOMICAL TIDE (L. A. T.). HIGHEST ASTRONOMICAL TIDE (H. A. T.). The lowest and highest levels, respectively, which can be predicted to occur under average meteorological conditions and under any combination of astronomical conditions. These levels will not be reached every year. These levels are not the extreme levels that can be reached, as storm surges may cause higher and lower levels to occur.

CHART DATUM. The actual levels vary widely at Standard Ports, and do not conform to any uniform tidal level. Modern practice is to establish datum at or near L. A. T. and datums are being gradually adjusted as opportunity offers so as to approximate L. A. T.

TIDAL PREDICTION

The problem facing the navigator is to determine the actual height of the tide above chart datum, or the time when there will be a preselected depth at a given place.

The method of tidal prediction has undergone a radical change and the latest method will be incorporated in the Ministry of Transport examinations in 1965.

The following pages show a reproduction of the prediction curves together with an explanation of the use in prediction at both Standard and Secondary Ports.

Each set of predictions for Standard Ports is accompanied by a diagram showing the mean tidal curves at both springs and neaps. The diagram for Glasgow is reproduced for the purposes of explanation. The curve for springs is the normal curve for a tide having a range of 13.5' and the curve for neaps corresponds to a range of 7.7'. The arguments used are the interval from H. W. and a factor. The curve shown on page 118 is for use at ports where Spring and Neap curves are not published.

The 'factor', when multiplied by the range, gives the height of the tide above low water. (Thus, the factor is 0 at low water and 1 at high water). (The A. T. T. publish a multiplication table to help the less arithmetically minded.)

For ranges at or greater than the spring range, the spring curve should be used, for ranges at or less than the neap range, the neap curve should be used. Where the range is at some intermediate value, use both curves and interpolate for the range of the tide on the day in question.

INSTRUCTIONS FOR FINDING HEIGHT OF TIDE AT TIMES BETWEEN HIGH AND LOW WATER.

1. Turn to the Standard Port predictions for the day in question.

2. Write down the heights of high and low water and, by subtraction, obtain the predicted range.

3. Compare the predicted range with the mean ranges given on the diagram and select the appropriate curve. (in doubtful cases, use both curves and interpolate).

4. Write down the predicted time of high water and the time for which the height is required and obtain the interval from H. W.

5. From the graph, read off the factor for the interval from high water.

6. Multiply the factor by the predicted range found in 2 above.

7. Add the answer to the predicted height of low water to obtain the predicted height at the time required.

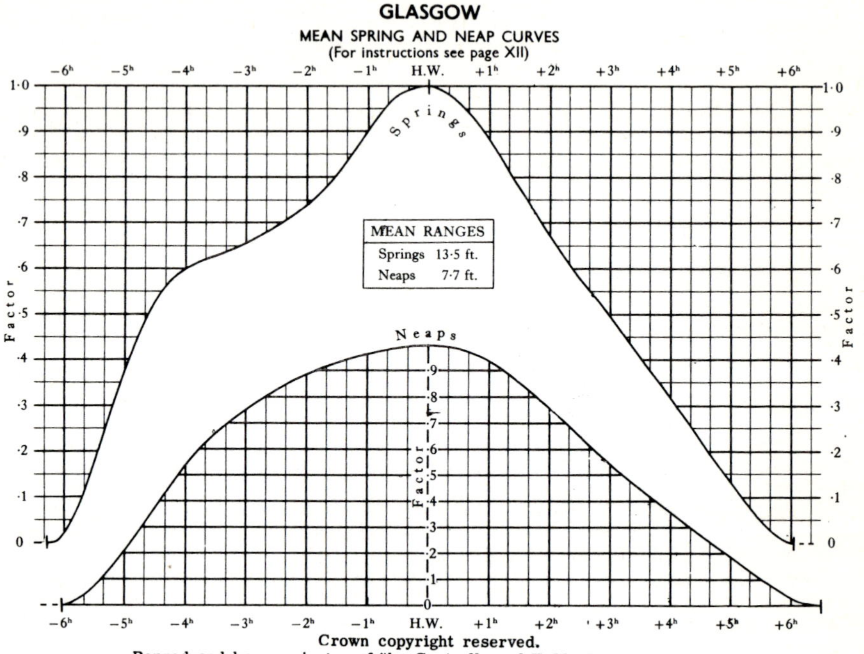

EXAMPLE 68

Find the height of the tide at 1600 on the 3rd February 1965 at Glasgow.

Extract from the A.T.T.

3 W	0154	12.2
	0753	1.5
	1406	13.8
	2014	0.2

H.W. height	13.8	H.W. time	1406
L.W. height	0.2	Time required	1600
Predicted range	13.6	Interval after H.W.	+0154

As range for the day is equal to the spring range, then this curve should be used.

From the spring curve

Factor for interval	+0154	0.70
Predicted range		13.6
Height above L.W.	(13.6 x 0.7)	9.5 (to first decimal)
Predicted L.W.		0.2
Predicted height at 1600		9.7

EXAMPLE 69.

Find the depth of water over a 1 fathom patch at 0645 on the 26th April 1965 Glasgow.

Extract from the A.T.T.

26 M	0256	4.4
	0843	9.8
	1528	2.7
	2143	10.0

H.W. height	9.8	H.W. time	0843
L.W. height	4.4	Time required	0645
Predicted range	5.4	Interval before H.W.	-0158

As predicted range is less than the neap range, the neap curve is used.

From the neap curve

Factor for interval	- 0158	0.88
Predicted range		5.4
Height above L.W. (5.4 x 0.88)		4.8
Predicted L.W.		4.4
Predicted height at 0645		9.2
Charted depth		6.0
Total depth at 0645		15.2

EXAMPLE 70

Find the a.m. time when the tide will have fallen to 10 feet at Glasgow on the 4th January 1965.

Extract from the A.T.T.

4 M	0135	12.9
	0739	2.2
	1355	14.2
	1956	1.2

Required height	10.0	Predicted height H.W.	12.9
Predicted L.W. height	2.2	Predicted height L.W.	2.2
Height above L.W.	7.8	Predicted range	10.7
Predicted range	10.7		

Factor $\left(\frac{\text{Height above L.W.}}{\text{Predicted range}}\right)$ 0.73

From spring curve with factor 0.73, interval after H.W. +0145
From neap curve with factor 0.73, interval after H.W. +0205

Interpolating for a predicted range of 10.7:-

Interval after H.W.	+ 0155
Predicted time H.W.	0135
Time when tide will have fallen to 10'	0330

PREDICTIONS AT SECONDARY EUROPEAN PORTS.

The tidal curve for a Secondary Port will seldom be exactly the same as the equivalent Standard Port. Nevertheless, except for certain areas (Swanage to Nab Tower) the curves for the Standard Port can be used to obtain predicted heights and times at the Secondary Port with a fair degree of accuracy.

ROSYTH

MEAN SPRING AND NEAP CURVES
(For instructions see page XII)

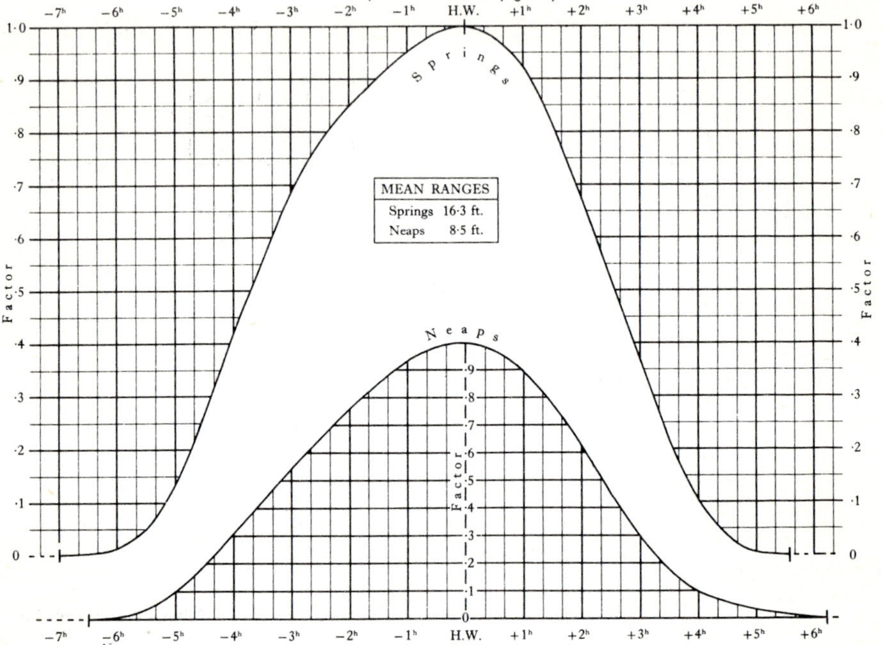

EXAMPLE 71

Find the height of the tide at 0500 on the 3rd January 1965 at Kincardine.

Extract from A.T.T. for Standard Port Rosyth:

3	0301	15.8
	0816	3.4
Su	1509	16.0
	2045	3.0

No.	PLACE	POSITION		TIMES (IN G.M.T.) AT STANDARD PORT				HEIGHTS (IN FEET) AT STANDARD PORT			
	STANDARD PORT	Lat. N.	Long. W.	High Water at 0300 and 1500	High Water at 1000 and 2200	Low Water at 0900 and 2100	Low Water at 0300 and 1500	MHWS	MHWN	MLWS	MLWN
227	**ROSYTH** . . . (see page 66)	56 01	3 27					17·7	14·0	1·3	5·8
	River Forth										
228	Grangemouth . .	56 02	3 39	+0 25	+0 10	−0 15	−0 52	+0·7	+0·4	0	+0·1
229	Kincardine . . .	56 04	3 43	+0 15	+0 30	−0 30	−0 30	+1·4	+0·8	+0·3	−0·2
229a	Alloa	56 07	3 47	+0 40	+0 40	+0 25	+0 25	+0·6	−0·1	−0·9	◈
229b	Stirling . . .	56 07	3 56	+1 01	+1 11	+4 38	+4 38	−9·1	−9·8	−2·0	−7·2

	H.W.height	L.W.height	H.W.time	L.W.time
Rosyth	15.8	3.4	0301	0816
Kincardine differences	+1.1	+0.1	+ 15	- 30
Kincardine predictions	16.9	3.5	0316	0746

Note: The height differences are obtained by interpolation or extrapolation of columns 9 and 10 for H.W. and columns 11 and 12 for L.W. The time differences are obtained by interpolation of columns 5 and 6 for H.W. and columns 7 and 8 for L.W.

	M.H.W.S.	M.H.W.N.	M.L.W.S.	M.L.W.N.
Rosyth	17.7	14.0	1.3	5.8
Kincardine differences	+1.4	+0.8	+0.3	-0.2
Kincardine	19.1	14.8	1.6	5.6

Kincardine M.H.W.S.	19.1	Kincardine M.H.W.N.	14.8
Kincardine M.L.W.S.	1.6	Kincardine M.L.W.N.	5.6
Kincardine spring range	17.5	Kincardine neap range	9.2
Kincardine H.W. height	16.9	Kincardine H.W. time	0316
Kincardine L.W.height	3.5	Time required	0500
Predicted range	13.4	Interval after H.W.	+ 0144
From spring curve Rosyth		From neap curve Rosyth	
Factor	0.75	Factor	0.71
Predicted range	13.4	Predicted range	13.4
Height above L.W.	10.1	Height above L.W.	9.5
Predicted height L.W.	3.5	Predicted height L.W.	3.5
Height at 0500	13.6	Height at 0500	13.0

Interpolating for a predicted range of 13.4:-

Height of tide at 0500 = 13.3 feet.

TIDAL PREDICTIONS FOR THE PACIFIC AREA.

In this case the Standard Ports do not have individual Spring and Neap curves and the prediction of intermediate times and heights is achieved using the next figure. This method is based on the assumption that the rise and fall of the tide takes the shape of a simple cosine curve. The principle of this method is the same as in the previous examples, where a predicted range is multiplied by a factor, the result being added to the height of low water to give the required height.

The difference from the previous graph is that allowance is made for variations in the duration of rise or fall by a series of curves at half-hourly intervals. The durations covered are from five to seven hours inclusive. For intermediate durations, interpolation should be employed but no extrapolation should be attempted.

(NOTE Should the duration of rise or fall be outside the limits of the curve, the correct method of prediction is to use the Admiralty Method of Prediction on form HD. 289).

INSTRUCTIONS.

1. Turn to the Standard Port predictions for the day in question.

2. Write down the heights of high and low water and, by subtraction, obtain the predicted range.

3. Determine the duration of rise or fall from the difference between the times of high and low water.

4. Determine the interval from high water for the time required.

5. From the diagram, using interval from high water and duration of rise or fall obtain the factor.

6. Multiply the predicted range by the factor.

7. Add the answer to the predicted height of low water to obtain the predicted height at the time required.

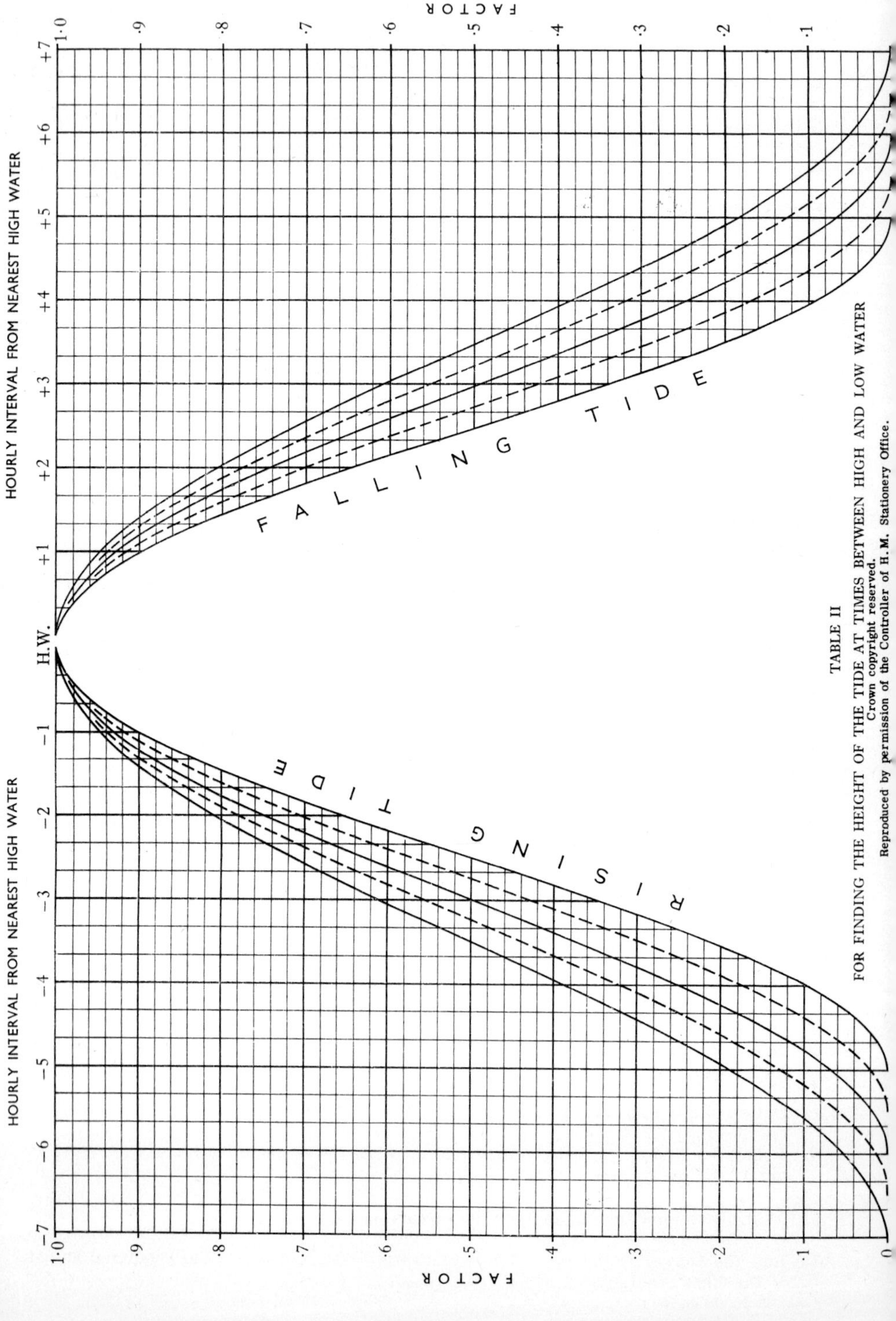

TABLE II

FOR FINDING THE HEIGHT OF THE TIDE AT TIMES BETWEEN HIGH AND LOW WATER

EXAMPLE 72

What is the correction to apply to the charted height of a lighthouse at Prince Rupert B.C. at 1900 on the 3rd April 1965?

Extract from A.T.T. Vol.III

3	0205	21.4
	0824	2.6
Sa	1432	20.4
	2032	4.2

Predicted height H.W.	20.4	Predicted time H.W.	1432	Time H.W.	1432
Predicted height L.W.	4.2	Time required	1900	Time L.W.	2032
Predicted range	16.2	Interval after H.W.	+0428	Dur.of fall	0600

Entering curve with duration of fall 6 hours and interval + 0428:

	Factor	0.15
	Predicted range	16.2
	Height above L.W.(16.2 x 0.15)	2.4 (to 1st decimal)
	Predicted height L.W.	4.2
	Predicted height at 1900	6.6
From Table V	M.H.W.S. Prince Rupert	21.2
	Correction to charted height	+14.6

EXAMPLE 73

At what time on the a.m. rising tide on the 4th June 1965 will there be 17 feet of water over chart datum at Prince Rupert B.C. ?

Extract from A.T.T. Vol.III

3	0324	22.0
	1010	1.5
Th	1638	19.0
	2218	7.9
4	0418	20.8
	1104	2.5
F	1736	18.5
	2321	8.4

Required height	17.0	Predicted height H.W.	20.8
Predicted height L.W.	7.9	Predicted height L.W.	7.9
Height above L.W.	9.1	Predicted range	12.9

$$\text{Factor } \left(\frac{\text{Height above L.W.}}{\text{Predicted range}}\right) = \frac{9.1}{12.9} = 0.71$$

Predicted time L.W.	2218 (3rd)
Predicted time H.W.	0418 (4th)
Duration of rise	0600

Entering curve with factor 0.71 and duration of rise 6 hours:

Interval before H.W.	-0210
Predicted time H.W.	0418
Predicted time for 17'	0208

TIDAL PREDICTIONS AT PACIFIC SECONDARY PORTS

All tides have semi-diurnal and diurnal components, and the latter causes inequality in successive heights and times of high and low waters.

When this inequality reaches certain limits, the A.T.T. publish average values of higher and lower high waters. (H.H.W. and L.H.W.) and higher and lower low waters (H.L.W. and L.L.W.) rather than mean spring and neap values (H.W.S., H.W.N., L.W.S., L.W.N.)

When predicting times of high and low water at a Secondary Port, the time differences from Part II are applied to the predicted times at the Standard Port on the day in question. If the Secondary Port has a small diurnal component the time differences can be applied to both high waters and both low waters without appreciable error. (see example 75) i.e. the time differences are for M.H.W. and M.L.W. Where the diurnal component is large the differences are for H.H.W. and L.L.W. These may be used for the time of L.H.W. and H.L.W. but the resultant times at the Secondary Port should be treated with caution as the probable error may be large. (see example 74)

In order to obtain the corrected values of the predicted heights at the Standard Port it is necessary to subtract the seasonal changes in mean level at the Standard Port. These are given at the foot of the right hand pages of Part II.

It is important to remember the algebraic rule of signs when applying these values

e.g.	Standard Port Prediction	15'.3
	Seasonal Change	-0'.6
	15'.3 - (- 0'.6) = 15'.3 + 0'.6 =	15'.9 corrected level
	Standard Port prediction	15'.4
	Seasonal Change	+0'.3
∴	15'.4 - (+ 0'.3) = 15'.4 - 0'.3 =	15'.1 corrected level.

Having obtained the corrected heights at the Standard Port it is now necessary to interpolate or extrapolate from the heights given for the Standard Port in Part II, a height difference for the Secondary Port.

After applying these height differences according to their sign the result is the uncorrected heights at the Secondary Port. The seasonal change in mean level at the Secondary Port is then added to these uncorrected heights to give the predicted heights at the Secondary Port.

e.g.	Uncorrected height at Secondary	14'.7
	Seasonal Change	+0'.6
∴	14'.7 + 0'.6 =	15'.3 predicted height at Secondary
	Uncorrected height at Secondary	-2'.4
	Seasonal Change	-1'.5
∴	-2'.4 + (- 1'.5) = -2'.4 - 1'.5 =	- 3'.9 predicted height.

EXAMPLE **74**

Find the times and heights of high and low tides on the 5th March at Lucena (Philippine Islands).

Extract for Standard Port Sandakan:

4 Th	0552	0.4
	1252	4.0
	1735	1.9
	2357	6.3
5 F	0619	0.7
	1259	4.2
	1808	1.6

SEASONAL CHANGES IN MEAN LEVEL

No.	Jan. 1	Feb. 1	Mar. 1	Apr. 1	May 1	Jun. 1	Jul. 1	Aug. 1	Sep. 1
4983, 4984	−0·3	−0·4	−0·3	−0·2	−0·1	+0·1	+0·2	+0·3	+c
4985–5013	−0·3	−0·4	−0·4	−0·3	−0·1	0·0	+0·2	+0·4	+
5014–5028	−0·3	−0·3	−0·3	−0·1	0·0	+0·1	+0·2	+0·3	
5115	+0·4	0·0	−0·4	−0·4	−0·2	−0·1	−0·1	−0·2	
5137							Assume nil		
7716	−0·4	−0·5	−0·5	−0·3	−0·2	0·0	+0·3	+'	

(89118)

No.	PLACE	POSITION		TIME DIFFERENCES		MEAN HEIGHTS (IN FEET)				Reference
	STANDARD PORT	Lat.	Long.			HHW	LHW	LLW	HLW	
5115	**SANDAKAN**	see page 26		HHW	LLW	6·3	4·0	1·3	2·7	
	SECONDARY PORTS	N.	E.	h. m.	h. m.	HEIGHT DIFFERENCES				
	Philippine Islands			(Zone	−0800)					
	Luzon									
4983	Lucena	13°54′	121°36′	+0013	+0010	−1·7	−1·4	−1·0	−0·9	

	H.H.W.	L.H.W.	L.L.W.	H.L.W.
Predicted times at Standard Port	2357 (4th)	1259	0619	1808
Time differences at Sec. Port	+ 13	+ 13	+ 10	+ 10
Predicted times at Sec. Port	0010 (5th)	(1312)	0629	(1818)

Times in brackets are to be treated with reserve.

	H.H.W.	L.H.W.	L.L.W.	H.L.W.
Predicted heights at Standard Port	6.3	4.2	0.7	1.6
Seasonal change of M.L. at Standard Port	-0.4	-0.4	-0.4	-0.4
Predicted heights less seasonal change	6.7	4.6	1.1	2.0
Interpolated height differences at Sec.Port	-1.8	-1.5	-1.0	0.9
Uncorrected heights at Secondary Port	4.9	3.1	0.1	1.1
Seasonal change of M.L. at Sec. Port	-0.3	-0.3	-0.3	-0.3
Predicted heights at Secondary Port	4.6	2.8	-0.2	0.8

Note: If intermediate times and heights are required at a Secondary Port the Admiralty method of Tidal Prediction on form H.D.289 must be used.

EXAMPLE 75

Find the times of the high and low tides at Bakapit on the 21st March 1965.

SEASONAL CHANGES IN MEAN LEVEL

No.	Jan. 1	Feb. 1	Mar. 1	Apr. 1	May 1	Jun. 1	Jul. 1	Aug. 1
5062	−0·2	−0·3	−0·2	0·0	−0·1	+0·1	+0·1	+0·
5097–5113	+0·2	0·0	−0·1	−0·2	−0·2	−0·2	−0·1	−0
5115	−0·4	[illegible]	−0·4	−0·4	−0·2	−0·1	−0·1	−(
5118–5122	+0·2	0·	−0·1	−0·2	−0·2	−0·2	−0·1	−
5125–5141							Assume nil	
5146–5150	+0·2	+0·1	−0·1	−0·3	−0·3	−0·2	0·0	
5233	+0·6	−0·6	+0·6	+0·6	−0·5	+0·1	−0·	
7021	0·0	−0·3	−0·4	−0·4	−0·2	−0·1		

Extract from Standard Port (Sungei Kutei):

21	0139	1.5
	0751	9.2
Su	1403	1.8
	1959	8.1

No.	PLACE	POSITION		TIME DIFFERENCES		MEAN HEIGHTS (IN FEET)				Reference
	STANDARD PORT	Lat.	Long.			HWS	HWN	LWS	LWN	
5233	**SUNGEI KUTEI**	see page 44		MHW	MLW	8·4	5·4	1·2	3·7	
	Darvel bay									
5103	Semporna	4 29	118 37	−0010	+0010	−2·6	−1·3	−0·6	−1·4	
5105	Kunak	4 41	118 15	−0005	−0005	−2·4	−1·5	−0·7	−1·1	t
5107	Lahad Datu	5 02	118 20	+0002	+0002	−2·7	−1·2	−0·5	−1·5	
5108	Bakapit	4 57	118 35	+0012	−0012	−2·4	−1·5	−0·6	−1·0	

	H.W.	H.W.	L.W.	L.W.
Predicted times at Standard Port	0751	1959	0139	1403
Time differences at Secondary Port	+ 12	+ 12	+ 12	+ 12
Predicted times at Secondary Port	0803	2011	0151	1415

Predicted heights at Standard Port	9.2	8.1	1.5	1.8
Seasonal change of M.L.at Standard Port	+0.6	+0.6	+0.6	+0.6
Predicted heights less seasonal change	8.6	7.5	0.9	1.2
Height differences at Secondary Port	-2.5	-2.1	-0.5	-0.6
Uncorrected heights at Secondary Port	6.1	5.4	0.4	0.6
Seasonal change of M.L. at Secondary Port	-0.2	-0.2	-0.2	-0.2
Predicted heights at Secondary Port	5.9	5.2	0.2	0.4

XII
REVISION QUESTIONS

MISCELLANEOUS QUESTIONS

1. Chronometer A is 8m. 20s. slow on chronometer B. B is 20m. 12s. fast on C. If chronometer C is 15m. 25s. slow on G. M. T., what is the error of A on G. M. T. ?

2. Establish a relationship between dep., dlong., and latitude.

3. The Sun rose bearing 107°C. and set bearing 247°C. Assuming no change of ship's head find the error.

4. Given that the radius of the Earth is 3960 miles and that the Moon's H. P. is $58'.0$. Calculate the distance separating Earth and Moon.

5. What is refraction and what effect has temperature upon refraction?

6. On a particular Mercator chart 1° of longitude measures $4''.2$. What distance would a measurement of $8''.4$ represent along the parallel of $51^{\circ}11'.0$?

7. The true altitudes of the upper and lower transits of a star were $77^{\circ}58'.0$ and $18^{\circ}08'.0$, respectively, both bearings were 180°. Calculate the latitude and declination.

8. Find, by means of the Traverse Table the distance steamed while taking a 2 point and 5 point bearing of a light which is 5 miles off when abeam.

9. What is the correction applied to D. F. bearings and why is it used?

10. On 1st March, chronometer B was 1m. 11s. slow on chronometer A. At the same time G. M. T. on 29th March, B was 2m. 22s. slow on A. If the daily rate of A was 2. 0s. gaining, find the daily rate of B.

11. From the following information plot the observed position. D. R. position $30^{\circ}19'.0$S., $24^{\circ}10'.0$W., azimuth of Sun 060°, intercept $11'.5$ towards. Vessel then steered 090°T. for 41 miles. An ex-meridian of the Sun bearing 010° gave a latitude of $30^{\circ}25'.0$S.

12. If the distance of the centre of the Moon to the centre of the Earth is equal to 60 times the radius of the Earth, calculate the H. P.

13. Describe the four adjustments of the marine sextant.

14. Explain the following:
    a. what is the vertex of a great circle?
    b. how far is the vertex removed from the equator.
    c. what is the great circle course equal to when crossing the equator?

15. Chronometer A is losing 2. 0s. daily and chronometer B is gaining 3. 2s. daily. If A is 10m. 15s. fast on B, what will the difference be in 20 days?

16. Describe a gnomonic chart. What is its principal use to the navigator?

17. Two vessels are 28 miles apart in latitude 44°S. If both steam due north keeping E/W from each other until they are 37 miles apart, calculate the difference of latitude made.

18. The L.H.A. of a star is $290^{\circ}07'.0$ and its declination is $12^{\circ}18'.0$S. If the latitude of the observer is $28^{\circ}21'.0$N., find the true azimuth.

19. Define the following terms: spring rise, neap rise, spring range and neap range.

20. A beacon of height 45 feet subtends a vertical angle of $1^{\circ}10'.0$. After steaming a distance of 3 cables it was abeam. What angle will it subtend after steaming a further $2\frac{1}{2}$ cables on the same course?

21. From the following information determine the geographical position of the Sun. Latitude 40°N., longitude 40°W., declination 10°N., L.H.A.T.S. 30°.

22. With the relevant information from the previous question find the geographical position of a star whose L.H.A. is $337^{\circ}30'.0$ and declination 20°S.

23. On 20th October, chronometer A is 4m. 01s. fast on G.M.T. and is losing 2.5s. daily. Chronometer B is 1m. 03s. slow on A. If B is 0m. 17s. fast on A on 30th November find the daily rate of B.

24. A vessel in position latitude $30^{\circ}05'.0$N., longitude $47^{\circ}30'.0$W. steers a course of 132°C. at 14 knots. After steaming for 3 hours she receives an S.O.S. from position latitude $30^{\circ}00'.0$N., longitude $47^{\circ}31'.0$W. Find the true course to steer to this position and if the speed is increased to 15 knots find the time taken to reach the position. (Variation 24°W. deviation 3°W.)

25. Calculate the D.M.P. between the parallels of 40°N. and 50°N.

26. In D.R. latitude $10^{\circ}54'.0$N., observed longitude $179^{\circ}54'.0$E. position line $160^{\circ}/340^{\circ}$, a vessel steered 040°T. for a distance of 44 miles. A meridian altitude gave the observed latitude as $11^{\circ}24'.0$N. Find the observed position by means of plotting.

27. Find two latitudes in which a star having a declination of $68^{\circ}51'.0$S. will bear 180° with a true altitude of $11^{\circ}50'.0$.

28. If twilight lasts until the Sun is 18° below the horizon, find the latitude where there will be twilight all night on one night only when the declination of the Sun is $23^{\circ}20'.0$N.

29. What is meant by the augmentation of the Moon's semi-diameter?

30. Show that D.M.P. x dep. = dlong x dlat.

31. What is the zenith distance of the Sun when its L.L. touches the horizon on 27th September 1958?

32. Show, by means of a figure, how the latitude may be determined by observing the upper and lower transits of an unknown star. How would you determine the declination of star?

33. Define the following: statute mile, geographical mile and nautical mile. Why is the latter not a constant quantity?

34. Draw a figure to illustrate the difference between horizontal parallax and parallax in altitude of the Moon. What is the connection between them?

35. Draw a figure on the plane of the observer's meridian and show the following: north celestial pole, celestial equator, ecliptic, a declination parallel of 10°N.

36. A triangle ABC is right angled at B. BA is 7.1 miles and BC is 11.36 miles. By means of the Traverse Table find the perpendicular from B to AC.

37. Show how 'mean latitude' may be converted to 'middle latitude'.

38. Illustrate and define the following terms:

observer's zenith
visible horizon
sensible horizon
rational horizon
dip
observed, apparent and true altitudes.

39. What is the principle of the construction of the Mercator chart and what is its chief advantage to the navigator?

40. Chronometer A is 32m. 05s. fast on G. M. T. at 1200 G. M. T. 1st May. On the same day chronometer B is 10m. 30s. slow on A. At 1200 G. M. T. 31st May A is 29m. 15s. fast on G. M. T. and B is 9m. 03s. slow on A. Find the errors of both chronometers on G. M. T. at 1200 G. M. T. 10th June.

41. Draw a figure to show a circumpolar star in latitude 50°N. Name some stars that would be circumpolar in this latitude.

42. Show that sin (parallax in altitude) = sin (H. P.). cos apparent altitude.

43. If the declination of the Sun is 17°06'.0 and its R. A. is 3h. 00m. 46s. find the obliquity of the ecliptic.

44. From the following information determine the vessel's position at 1100 by means of plotting:

0900 D.R. latitude 51°10'.0N., observed longitude 10°14'.0W. az. 220°T. course 090°T., speed 12 knots.
1100 Fastnet light (51°23'.0N., 9°36'.0W.) bearing 070°T.

45. Calculate the difference in velocity of the Earth's surface at the parallels of 25° and 48°.

46. A vessel A is on the equator and steering due east at 16 knots.
Another vessel B is steering due west along a parallel of latitude at 12 knots. If A makes a d'long of 60' while B makes a d'long of 48' find the latitude of B.

47. From latitude 59°40'.0N. a vessel steamed south until the mean latitude was twice the d'lat. Find the latitude reached.

48. While steering a course of 060°T. a vessel made a d'lat of 31' and a dlong of 66'. Find the latitude reached and left.

49. A vessel's course is 042°T. and makes a departure of 200 miles and D. M. P. 240'. Find the latitude left.

50. Two vessels are on the same parallel of latitude, but separated by 2°41'.0 of longitude. A steers 030°T. and B steers 330°, both vessels making the same speed. If the latitude left was 31°00'N. find the latitude reached when they are 10 miles apart.

51. A vessel in latitude 57°N. steers 300°T. at 24 knots. Find the rate of change of longitude per hour.

52. Two vessels are on the same meridian, vessel A is on the equator and steers 270°T. Vessel B is to the south of A and steers due west maintaining the same longitude as A. If the speed of B is 3/4 that of A, find the latitude of B.

53. A vessel in position latitude 55°40!0N., longitude 15°00!0W. set course to reach position latitude 56°08!0N., longitude 15°00!0W. After steaming for 28 miles it was found that the compass error had been applied the wrong way, the vessel now being in position latitude 56°07!8 N., longitude 14°54!0W. What is the correct compass error?

54. Two stars bearing 140°T. and 240°T. gave an observed position as latitude 30°10!0N., longitude 35°10!0W. Afterwards it was found that the index error of $1\frac{1}{2}'$ off the arc had not been applied. Find the true position.

55. An observer found that by using a D.R. latitude of 51°55!0N. the observed longitude was 20°04!0W. and by using latitude 52°05!0N. the observed longitude was 19°54!5W. Calculate the azimuth of the body.

56. Define the following terms in relationship to tides: priming
lagging

57. Star sights gave a position of latitude 56°01!0N., longitude 7°06!0W., the error of the chronometer being assumed as 12m. 42s. slow. After steering 070° for 11 miles, the vessel found that a light in position latitude 56°08!0N longitude 6°38!0W. bore 050°T. distance 5 miles. Find the true chronometer error.

58. If the ratio of the longest day to the shortest day is 3 to 1. Find the latitude if the obliquity of the ecliptic is 23°26!5.

59. On the longest day of the year, the Sun's centre just touches the rational horizon below the pole. Find the latitude.

60. Define the following terms:

apogee, perigee, conjunction, opposition and quadrature.

61. From the information given in example 52 obtain the intercept and the position through which to draw the position line.

62. From the information given in example 53 obtain the intercept and the position through which to draw the position line.

63. From the information given in example 55 obtain the position through which to draw the position line by "longitude by chronometer".

64. As question 63 but using the information given in example 54.

65. Obtain the intercept and hence the position through which to draw the posi-line from the information in example 56.

66. As question 65 but using the information in example 57.

67. From the information given in Example 71 determine the earliest time after 0900 there will be 20 feet of water over a lock sill 10 feet below chart datum at Grangemouth on 3rd January 1965.

68. From the information given in Example 71 find the height of the tide at Grangemouth at 1200 on 3rd January 1965.

69. From the information given in Example 72 find the time in the first dog watch when there will be 15 feet of water above chart datum at Prince Rupert B.C. on the 3rd April 1965.

70. From the information given in Example 73 find the correction to the lead-line depth at 0600 on the 3rd June 1965 at Prince Rupert B.C.

NUMERICAL ANSWERS TO REVISION QUESTIONS

1. 3m.33s. slow.
3. 3° east.
4. 234,726 miles.
6. 75.25 miles.
7. latitude $48^\circ 03'.0$S. declination $60^\circ 05'.0$S.
8. 8.7 miles.
10. 0.54s. losing.
11. latitude $30^\circ 27'$S. longitude $23^\circ 03'.0$W.
12. H.P. $57'.3$
15. 8m.31s.
17. $1554\frac{1}{4}$ miles.
18. $110^\circ.7$
20. $1^\circ 18' 41''.5$
21. Sun's geographical position - latitude 10°N., longitude 70°W.
22. Star's geographical position - latitude 20°S., longitude $17^\circ 30'$W.
23. 0.55s. losing.
24. 278°T. 2 hours 48 minutes.
25. 848.9
26. latitude $11^\circ 24'.0$N. longitude $179^\circ 35'.5$W.
27. latitudes $9^\circ 19'.0$N. and $32^\circ 59'.0$S.
28. latitude $48^\circ 40'.0$N.
31. $90^\circ 18'.4$.
36. 6.02 miles.
40. A is 28m.18.4s. and B 19m.43.4s. both fast on G.M.T.
43. $23^\circ 26'.8$
44. latitude $51^\circ 19'$N. longitude $9^\circ 53'.3$W.
45. 213.46 knots.
46. latitude $20^\circ 22'.0$N.
47. latitude $35^\circ 48'.0$N.
48. latitude $35^\circ 48.4$N and $35^\circ 17'.4$N.
49. latitude $20^\circ 24'.4$N.
50. latitude $32^\circ 50'.9$N.
51. $38'.16$ per hour.
52. latitude $41^\circ 24'.2$S.
53. $3\frac{1}{2}^\circ$ east.
54. latitude $30^\circ 07'.9$N., longitude $35^\circ 10'.3$W.
55. $120^\circ.5$ or $300^\circ.5$.
57. 12m.30.0s. slow.
58. $58^\circ 29'.0$
59. latitude $66^\circ 33'.5$N or S. which equals polar distance of Sun's centre.
61. Intercept $3'.6$ towards.
    P.L. $123^\circ.2$ - $303^\circ.2$ through latitude $10^\circ 57'.0$N. longitude $109^\circ 28'.0$E.
62. Intercept $1'.2$ away
    P.L. $059^\circ.4$ - $239^\circ.4$ through latitude $52^\circ 04'.03$N., longitude $20^\circ 20'.99$W.
63. P.L. $050^\circ.5$ - $230^\circ.5$ through latitude $40^\circ 12'.0$S., longitude $23^\circ 45'.0$W.
64. P.L. $138^\circ.9$ - $318^\circ.9$ through latitude $40^\circ 04'.0$S., longitude $100^\circ 02'.5$W.
65. Intercept $1'.0$ away.
    P.L. $094^\circ.6$ - $274^\circ.6$ through latitude $27^\circ 03'.0$N., longitude $16^\circ 14'.9$W.
66. Intercept $1'.0$ away.
    P.L. $103^\circ.2$ - $283^\circ.2$ through latitude $31^\circ 21'.0$S., longitude $155^\circ 39'.8$E.
67. Earliest time 1150.
68. Height of tide 9.6 feet.
69. Time required 1650.
70. Correction to leadline -15.4 feet.

INDEX

Printed by Westwood Trading Co. Ltd. South Croydon, Surrey.